Thicker Than Water

Thicker Than Water

A Social and Evolutionary Study of Iron Deficiency in Women

Elizabeth M. Miller

OXFORD UNIVERSITY PRESS

OXFORD
UNIVERSITY PRESS

Oxford University Press is a department of the University of Oxford. It furthers the University's objective of excellence in research, scholarship, and education by publishing worldwide. Oxford is a registered trade mark of Oxford University Press in the UK and certain other countries.

Published in the United States of America by Oxford University Press
198 Madison Avenue, New York, NY 10016, United States of America.

Library of Congress Cataloging-in-Publication Data
Names: Miller, Elizabeth M. (Elizabeth Marie), 1981– author.
Title: Thicker than water : a social and evolutionary study of
iron deficiency in women / by Elizabeth M. Miller.
Description: New York, NY : Oxford University Press, [2023] |
Includes bibliographical references and index.
Identifiers: LCCN 2023011400 (print) | LCCN 2023011401 (ebook) |
ISBN 9780197665718 (hardback) | ISBN 9780197665732 (epub) | ISBN 9780197665749
Subjects: MESH: Iron Deficiencies | Women's Health | Gender Equity
Classification: LCC RA564.85 (print) | LCC RA564.85 (ebook) |
NLM QU 260.5.I7 | DDC 613.2082—dc23/eng/20230510
LC record available at https://lccn.loc.gov/2023011400
LC ebook record available at https://lccn.loc.gov/2023011401

DOI: 10.1093/oso/9780197665718.001.0001

Printed by Sheridan Books, Inc., United States of America

Contents

Preface

I became interested in women's iron deficiency anemia while I was writing my dissertation. Charged by my doctoral committee to publish *something*, I dove into some of the data I had collected but would not be using in the dissertation itself. I had always been fascinated by the trade-offs women faced when having their babies. Not trade-offs in work, family, or hobbies; rather, evolved trade-offs that governed their physiology. Reproduce now, or save energy to reproduce later? Expend energy on pregnancy, or fight off an infection? Use bodily resources to feed your fetus, or nourish yourself? Armed with this perspective, I tested the trade-offs between women's iron status and reproduction, producing the specified paper in the *American Journal of Human Biology* in 2010. I continued to publish papers here and there on women's iron status from an evolutionary perspective, considering it a side project, until one day I recognized that I had barely scratched the surface of women's iron deficiency and neither had any other contemporary anthropologist. In addition to the evolved trade-offs that governed women's iron status, it was clear the social contexts of iron were both overwhelming and understudied. I soon had a blueprint for this book.

The goal of this book is to illuminate the topic of iron deficiency in women by integrating both social and evolutionary theories into a unified picture of human biology. Women are especially vulnerable to iron deficiency and/or anemia, contributing to thousands of maternal deaths per year. That this phenomenon has been overlooked by anthropologists is shocking, considering it is the consequence of evolved biology and massive global inequalities. The "easy" cure for iron deficiency—iron supplementation—seemed to be accepted without comment as the final word on the problem.

This book rectifies this omission. Iron deficiency is an ideal lens to study physiology, evolution, and society. It is a form of malnutrition, meaning that modern diets are partly to blame. It can demonstrate how social problems "get under the skin" and become biology. And because iron deficiency is an evolved vulnerability that is ultimately caused by social inequalities, it is a problem that needs social solutions—doling out iron pills is not enough.

This book is for those who are anthropologically curious and want to examine human biology from all angles. I take several theoretical perspectives in the book, derived from both evolutionary and social theory. I discuss how women's reproductive physiology evolved to make them vulnerable to iron deficiency—but that *vulnerable* does not mean *afflicted*. Iron deficiency only occurs on a worldwide scale because social conditions make it happen. I discuss how sexism, racism, and global inequalities become embodied as iron deficiency, causing millions of women to suffer, particularly during pregnancy and birth. And I discuss how iron is distributed across the world, both within the landscape and within food. The conditions that make environments iron deficient are no accident, but a consequence of unequal social structures that humans have built. Iron deficiency, then, is but one example of the deep enmeshment of the biological and the social and shows that the mosaic of human health allows for few easy solutions.

Acknowledgments

There are so many who have supported the journey of this book. I would particularly like to thank the women and infants who have participated in my projects over the years. Without you, I would understand very little about anthropology or iron. I appreciate your generosity more than you could know.

I have many colleagues who supported me while writing this book. My writing group, particularly Heather O'Leary, Daniel Lende, and Holly Donahue Singh, have consistently helped me set goals and celebrate milestones. Jim McKenna provided helpful feedback and encouragement on part of this work. Adetola Louis-Jacques has been a constant collaborator and I am thankful she is willing to entertain my ideas about iron and the microbiome. My work with Achsah Dorsey, Maie Khalil, and Alexis Monkhouse have been critical in shaping my views on iron. As always, my views are my own.

On a personal note, I could not have made it through this manuscript without friends, family, the little dog, and especially Andrew. Thank you, and I love you.

Acknowledgments

1 Ironing Out Evolutionary and Social Theory

Human bodies are a sea of matter. While we mostly consist of salt water—about 60% of our bodies are water—we also have organic matter and tiny amounts of inorganic "stuff" arranged in a mosaic of Life. That organic matter and that stuff help form the structure of the body and keep it running. Iron, one of the most common elements on Earth, is vital to the body's proper functioning. Without it, we would die. Iron is responsible for transporting oxygen to cells, which need oxygen to produce energy. Most iron in the body lives in red blood cells in the form of hemoglobin. The red blood cells circulate between the lungs and the far tissues of the body, keeping oxygen at an optimal level. Ideally, iron levels are optimized too—just enough to keep the oxygen flowing and no more.

When something goes wrong with this system—perhaps because there is not enough oxygen in the body—our first instinct is to call a doctor. After all, doctors are supposed to be the experts on the human body. Much of our understanding of iron in human bodies is attributed to biomedicine, or the scientific approach to doctoring. Biomedicine makes iron simple: sometimes the level of iron in the body is low. When it is, you supplement it, choking down large red pills until your iron level is higher. Supplemental iron makes its way to the red blood cells and becomes hemoglobin, which helps transmit oxygen to cells. This relieves low hemoglobin (anemia), and everything is fine.

Supplementation forms the basis for biomedical practice and public health policy for micronutrient deficiencies. The simplicity of supplementation—which appears to be just an obvious bit of math—means that no one should be suffering from iron deficiency or anemia as long as we can transport iron pills to the right places. Thus, iron deficiency merely becomes a question of logistics. Get the pills, put them in mouths. Nothing is simpler than that.

Why, then, do one third of reproductive aged women in the world suffer from anemia, at least half of which is caused by iron deficiency? And why does having iron deficiency anemia make them twice as likely to die during pregnancy, childbirth, and postpartum? Why don't men experience anemia

Thicker Than Water. Elizabeth M. Miller, Oxford University Press. © Oxford University Press 2023.
DOI: 10.1093/oso/9780197665718.003.0001

to the same degree? And why does this happen more to the most disadvantaged women, the women suffering poverty and discrimination? If ameliorating iron deficiency anemia were just as simple as getting pills into mouths, this problem would have been solved by now. Simplicity has turned women's iron deficiency into a false dilemma. A false dilemma comes about when a complex problem is presented as though it were a simple question of black and white. Biomedicine offers a quick, simple fix to iron deficiency—but the problem is much more complex than that.

We need more than biomedicine to understand the problem of women's iron deficiency. Did you know that poor women are more likely to be anemic than rich women? That means iron deficiency is social as well as biological, reflecting the social inequalities that make some people poor and others rich. Did you know that iron deficiency has had several different names over the centuries, each illness having a particular population and set of symptoms? This means that the concept we call "illness" is socially constructed, with a definition, cause, and cure that has changed over centuries. And did you know that women have evolved a set of physiological processes to help maintain their iron stores during pregnancy while protecting their developing offspring against iron's harmful effects? At least, they do if enough iron is available. Women's widespread iron deficiency happens because their evolved bodies interact with social realities. Both things, it turns out, are outside the typical scope of biomedical science. Luckily, other fields—like anthropology—have evolution and social science at the core of the discipline. That means anthropologists like me can look at iron deficiency in a way few have done before.

This book uses anthropological tools to understand the ebbs and flows of women's iron physiology. I'll piece together the evolution, history, society, and culture of iron, outlining what know, what we think we know, and what we were completely wrong about. Rather than uphold black and white explanations of biomedicine in understanding iron deficiency, I will demonstrate a colorful mosaic of explanations, ranging from biological to cultural to social, that will show why iron deficiency persists in women, and why women and iron are such a deadly combination. To do this, we first must put ourselves in the frame of mind of an anthropologist. And an anthropologist, primarily, knows how to ask the right questions.

The hows and the whys of asking questions

The late evolutionary biologist Theodosius Dobzhansky authored an essay entitled "Nothing in Biology Makes Sense Except in the Light of Evolution" in 1973. In it, he highlighted the need for evolution in biology education, arguing that seemingly disparate facts in biology can be linked together with evolutionary explanations (Dobzhansky 1973). This phrase became a rallying cry with biology educators—I use it myself in classes to highlight the importance of learning evolutionary theory. However, many students in the United States never hear it: evolution is frequently not taught in lower-level biology. A small, but noisy, number of religions see evolution as controversial, and it has therefore been deprioritized in educational practice. Indeed, even within higher education, evolutionary biology and cellular and molecular biology often inhabit separate academic departments, meaning that a host of budding biologists have little exposure to evolution. Given that well over half of incoming medical students in 2021 listed the biological sciences as their major (Association of American Medical Colleges 2021), this means that evolutionary theories are frequently deprioritized in the biomedical pipeline. This is recapitulated in medical school curriculums and ultimately affects biomedical research. Biomedical questions are rarely informed by evolutionary biology, preferring instead to answer the question: "How does that work?"

Asking "how" is crucial to biology research. How does DNA replicate? How does a plant transform the sun's energy into metabolic energy? How do you alleviate iron deficiency? These are all extremely important questions that can be converted into testable hypotheses. These testable hypotheses are how we've produced scientific advances that have improved lives.

Asking "why" these things happen is more enlightening to the evolutionary scholar. Asking "why" DNA replicates opens a host of new questions. Answers to the "whys" can be simple: because to make new cells, a cell needs to replicate DNA, and that is why DNA replicates. But follow the "whys" far enough, and you'll need evolution. Why does DNA replicate? Because cells, as the smallest unit of Life, evolved to replicate themselves. When you get to the evolutionary "why," you have a unique way of understanding the phenomenon at hand.

Evolutionary scientists tend to distinguish evolutionary questions from mechanistic questions. They call the evolutionary "why" questions *ultimate* questions, and mechanistic "how" questions *proximate* questions. While both are important, the proximate questions are given more weight in fields like biomedicine or molecular and cellular biology. As you might imagine, fields where you can experimentally manipulate conditions—exposing cells, mice, or people to different situations—tend to favor proximate questions. This includes fields such as microbiology, molecular and cellular biology, immunology, and biomedical science. These distinctions are not fixed: some very elegant experiments have been derived from ultimate questions, and observation is still frequently used for proximate questions. The most important part is the question itself, not the method used to answer it.

The question "how do we alleviate iron deficiency?" led to a simple answer. In the 17th century, an English physician by the name of Thomas Sydenham treated chlorosis, or anemia, to remarkable success. As a physician who relied on empiricism and observable signs of disease, he was suspicious of theory and believed that medicine should be learned from the bedside, not books (despite adhering to the Hippocratic concept of the four humors, of course). He gave raw iron—iron shavings—to patients with "Hysterick Diseases," of which chlorosis (anemia) was one, which tended to improve the chlorotic symptoms (Sloan 1987). While the mechanisms of how iron deficiency caused anemia and how iron supplementation could alleviate it would not be established until the late 19th and early 20th century, Sydenham nevertheless was able to use observation and hypothesis testing to establish an effective treatment. It's why this simple answer has managed to persist—and sometimes work—through the 21st century.

This simple answer still does not address many of the "whys"—why are women more vulnerable to iron deficiency anemia? Chlorosis, after all, was considered a woman's disease. Why is it especially harmful during pregnancy? And why, if we have the technology to stop it, is it so persistent? Evolution can help us provide the answer.

Evolution is . . .

There are a lot of misconceptions about evolution. "Evolution is just a theory," "Humans are descended from monkeys," or "Evolution is always progressing toward a goal." This stems, in part, from inadequate education in evolution. On the other hand, evolution is just difficult to wrap your mind around sometimes. The deep time needed for major evolutionary changes is not easy for the human mind to grasp. And it is easy to "think like a Lamarckian" by attributing intention, progress, and biological ideals to evolution. In this section, I'll give an overview of evolutionary history and theory and clear up common misconceptions about evolution.

The story of Charles Darwin's theory of natural selection is well known: young man takes a five-year journey around the globe; mulls ideas over in his head for decades until a man who had a similar theory of natural selection appear to him in a malaria-induced fever dream (Alfred Russel Wallace) asked him to co-present on the topic; whips up 200,000-ish-word abstract called *On the Origin of Species* outlining how life evolved via natural selection; and becomes a titan of science. Over 150 years later, this story still resonates. Even Darwin's self-doubts are relatable to scientists: "But I am very *poorly* today & very *stupid* & I hate everybody & everything. One lives only to make blunders," written to his friend, fellow scientist Charles Lyell (Darwin 1903). Darwin successfully nailed the "why"—explaining how change over time created enormous diversity of life. He was a little less clear on the "how," and science is still trying to catch up to all the ways that life evolves.

Darwin observed that living creatures seemed to "fit" their environments. He also observed that individuals in a population were naturally variable, meaning that some had characters that "fit" the environment better than others. The individuals with the better-fitting characteristics could be more successful in their environment, meaning that they were better able to survive and reproduce. Darwin hypothesized that these characters were heritable, meaning that there would be more offspring in the next generation with the better-fitting characters compared to offspring of those whose variants did not fit as well. Over time, the better-fitting traits would become more abundant in the population, and eventually this process could lead to new species. This was Darwin's theory of natural selection. He marshalled abundant evidence of natural selection in *On the Origin of Species*.

The way Darwin conceptualized evolution is slightly different than how it is taught today. For one, although he hypothesizes that characters are inherited, he does not elaborate on a mechanism for how they become inherited, nor explain how new characters are introduced into the population. At least, not in *On the Origin*—Darwin's later work shows that he did develop a hypothesis of heredity, called the Pangenesis hypothesis. In this hypothesis, small particles called gemmules circulated though the body, eventually resting in the gonads to be transmitted to offspring; if the environment modified gemmules, these modifications would be transmitted to offspring (Liu 2008). However, because Darwin's original conception of natural selection does not have a formalized mechanism of inheritance, it is different than what is taught in introductory classes today.

Heredity bloomed into its own field in the 19th century. August Weismann proposed the theory of the continuity of the germplasm, where he hypothesized that chromosomes within sex cells (eggs and sperm), and not somatic cells (body cells), were passed, intact, from parents to offspring. The germplasm was his name for hereditary material of the cell. He rejected the inheritance of acquired traits—the famous "Lamarckian" theory of heredity in which traits that are acquired during an organism's lifespan are transmitted to offspring. He had a remarkably interesting way of showing it. He removed the tails from mice and bred them over five generations. Since they did not transmit the acquired amputation, he declared the inheritance of acquired traits to be unsupported by evidence. Of course, this experiment had limitations that Weismann was well aware of, but he was responding to claims that animal mutilations could be inherited. He got the point across, anyway.

Francis Galton, noted statistician, social scientist, and eugenicist, played a Janus-faced role in the fight for the supremacy of heredity. On one hand, he established the field of statistics. On the other hand, he founded the field of eugenics, which is the study of manipulating human reproduction to promote "favorable" characteristics in the population. He disproved Darwin's Pangenesis theory of heredity and established his own law of ancestral heredity, arguing that on average, half of each parent contributes to the characteristics of their offspring. He established twin studies, adoption studies, and even invented correlation to understand heredity, focusing on the inheritance of abilities. By the early 20th century, the term "genetics" generally meant "heredity," meaning that observed characteristics—ranging from hair color to height to behavior and personality—were inherited from parents. Unfortunately, they were also considered markers of racial group membership. From there, it was a short step to espousing eugenics. Eugenics was not a minority viewpoint at the time: it was a central part of the field of genetics. Eugenics was a movement picked up by leading intellectuals, politicians, and policymakers. While the stated objectives were to promote the reproduction of people with "desirable" traits, in practice discrimination, forced sterilization, and even genocide were the tools of eugenicists. Unsurprisingly, racial minorities and the poor were the frequent target of eugenicist policies. The Holocaust, for example, was influenced by eugenicist viewpoints that were espoused by intellectuals in the United States.

The 20th-century rediscovery of Gregor Mendel's 1866 work on pea plants revolutionized genetics research. While the discovery did not overtake racialized beliefs in human genetics—instead, it likely fueled it—it did uncover mathematical consistencies in the heredity of certain characters. Mendel was a Catholic friar in the order of St. Augustine, living in St. Thomas's Abbey in what is now the city of Brno in the Czech Republic. His experiments with pea plant crosses established the modern science of genetics. He showed that a yellow pea from a line that always bred yellow crossed with a green pea from a line that always bred green produced offspring that were always yellow. When the yellow peas were crossed, their offspring were yellow and green in a predictable 3:1 ratio. Mendel hypothesized that each parent plant passed a sort of unseen "factor" to their offspring, and that some factors can override other factors in the expression of a trait.

Through these results, he established the concept of dominant (yellow) and recessive (green) traits, as well as the actions of invisible factors—what we now call genes and alleles—that determine traits in a predictable way. He also determined these factors segregate independently within the parents, meaning that parents have two factors determining each trait, and they pass one at random to their offspring; and that when you look at two or more traits, their factors assort independently from one another in parental gametes. Mendel published this work in the German-language journal *Proceedings of the Natural History Society of Brünn* and sent reprints to various scientists. However, Mendel's work went almost unremarked upon—cited only 3 times in 35 years—until rescued from obscurity by Hugo de Vries, Carl Correns, and Erich von Tschermak-Seysenegg in 1900, who separately confirmed Mendel's results; and William Bateson in 1909, who translated Mendel's article into English and gave his initial work a wider audience (Keynes and Cox 2008).

The nascent field of population genetics in the early 20th century relied on the mathematical consistencies of Mendelian inheritance. Population genetics is primarily concerned with how populations change over time by looking at how the proportions of alleles in a population vary over generations. G. H. Hardy and Wilhelm Weinberg independently developed equations to test whether a population's alleles were stable across generations in 1903, leading to the Hardy-Weinberg equilibrium taught in introductory biological anthropology classes. The population geneticists J. B. S. Haldane, Ronald Fisher, Sewall Wright, and Theodosius Dobzhansky mathematically modeled forces of evolution such as natural selection, genetic drift (change in character frequency across generations due to chance), and mutation (the addition of new characters to a population). As evolutionary mechanisms became enmeshed with Mendelian genetics via population genetics, a new way of thinking about evolution emerged. The term Modern Synthesis refers to the integration of population genetics with such diverse fields as geology, paleontology, botany, and systematics. The Modern Synthesis describes evolution at the population level (microevolution) and the species level (macroevolution).

Microevolution is, essentially, population genetics. It describes the change in allele frequencies in a randomly breeding population over generations. Alleles are defined as different forms of a gene (e.g., there is a gene for pea color with two alleles, yellow and green). If the allele frequencies of a population are not changing, the population is said to be in equilibrium. If they are, then evolution is happening; or there is a violation of population assumptions (such as non-random mating). There are four forces of evolution that affect populations: gene flow, genetic drift, mutation, and natural selection. Gene flow is the introduction of alleles from one population into another. Genetic drift is the change in frequency from generation to generation due to random sampling of parental alleles; the effects of genetic drift are stronger in small populations than larger ones. Mutation is the introduction of a brand-new allele into the population due to modification of an existing allele. And natural selection is, of course, the change in allele frequency due to a selective force, as Darwin described. Natural selection was believed to be the most powerful force of evolution. When natural selection occurs for a long enough time, populations could split into new species. Evolution at this level—the change in species over time—is macroevolution. Macroevolution takes place at incredibly long time scales—tens of thousands to billions of years—and is primarily concerned with species. Macroevolution focuses on how species are related to each other and how they are descended from ancestral species. The fossil record, which partially captures evolved changes, is a major source of data for understanding macroevolution.

The Modern Synthesis has long since shown its age. The structure of DNA, the "invisible factor" in which genes and their alleles rest, was discovered in 1953—*after* the coalescence of the Modern Synthesis—and the age of molecular genetics was born. Classical Mendelism fits pretty well onto the meiosis of DNA in gametes, especially the law of segregation; but the law of independent assortment is hampered when two genes exist on the same chromosome, or when they are linked in some other way. Most traits are not Mendelian either. Instead, they are frequently continuous, meaning that there are many more than two forms of a trait (think about height and all of the different possible shades of hair, skin, or eye color). Other aspects of modern biology are also given short shrift by the Modern Synthesis: modifications to DNA that can change how the DNA is expressed (epigenetics), gene expression, cellular mechanisms that modify how protein is expressed, growth and development of the body, physiological mechanisms, microbiomes that coexist with the body, behavior, society and culture, niche construction, and more.

These biological processes have more or less been integrated into an evolutionary framework—the Extended Evolutionary Synthesis—but since many of these mechanisms can be inherited in some fashion, it calls into question the primacy of Mendelism for our understanding of inheritance. And since many of these mechanisms can be modified by the environment, it hastens the timescale through which evolutionary modifications can be transmitted via generations. It also greatly expands the meaning of environment—organisms can modify and be modified by their environments, and their environments can include interactions with many other organisms. Other members of their species, individuals in other species, and species that coexist within and on their bodies are all part of an organism's environment. In humans, environments extend to human society and culture, which can have a profound impact on the functioning of our evolved bodies. Sometimes, they can even make us sick.

Dr Darwin

One morning, you wake up and feel a little off. You get dressed and head out, only to start shivering halfway through the day. You reach into your pocket and take two aspirins. Within half an hour you are no longer cold, but instead drenched in sweat. However, your fever is now gone. You continue going about your day until the medication wears off four hours later, and you begin shivering again. In the meantime, you've infected three other people with the flu.

When you take an over-the-counter medication to reduce a fever, it's easy to forget how the medication works. The hypothalamus, a small region near the base of the brain, acts as the body's thermometer. Normally, the hypothalamus regulates the body's temperature within a strict range. When the immune system interacts with an infection, it sends immunological signals called pyrogens to the hypothalamus. This signals to the hypothalamus that the body's temperature should be set several degrees higher than the normal range. The reason that fevers are accompanied by shivering is that it can be difficult for the body to raise its temperature high enough to satisfy the hypothalamus, and thus you feel too cold. When you take a medication for fever like acetaminophen or naproxen, it blocks the action of pyrogens. This lowers the temperature set point on the hypothalamus. When your brain is at a lower setting than your body, your body will start to sweat profusely as it seeks to rid itself of excess heat.

This is the "how" of fevers. But after suffering through a fever once, you may be wondering, why do we have them? At first, you may think that the infectious pathogen—the virus or bacteria—is responsible for releasing pyrogens with the goal of damaging the body. While infectious pathogens like bacteria and viruses can cause damage to the body, they aren't directly responsible for raising the body's temperature. Instead, the immune system releases the pyrogens in response to the infection. So why, then, does the body raise its own temperature? Wouldn't it cause less harm and fewer days out of work if the immune system responded differently?

The answer to this question comes from an unusual place: the desert iguana (*Dipsosaurus dorsalis*). Desert iguanas, unlike mammals, do not maintain a warm body temperature via a brain-based "set point" sustained by their metabolism. Instead, they regulate their body temperature behaviorally by moving between warmer and cooler places. In the 1970s, scientists began to wonder if lizards like the desert iguana had a fever response that was regulated by the brain.

To test this, researcher Matthew Kluger and colleagues injected the iguanas with bacteria and released them into simulated desert habitats. In one group, the lizards were given fever-suppressing medication and in the other, they were given no drugs. The iguanas in the fever group spent about five days in a warm part of the habitat before returning to cooler temperatures. They survived. In the group who were given fever-reducing medication, about half died of the infection. This study upended what scientists thought about the point of fever. Further experiments in other animal systems confirmed that fever reduced infectious disease mortality (Kluger et al. 1998). Fevers work, whether the hypothalamus raises the body's set point or an animal seeks out enough warmth.

Fevers are not a perfect solution to infection. There is risk of brain damage when running a fever over 107°F. However, as the desert lizards show, the risk of dying from infection generally outweighs the risk of brain damage from fevers. A lower, but still important, cost of running a fever is the energy it takes to keep the body at an elevated temperature. This cost may not seem like much, but fevers require significant energy if you regulate your own body temperature as mammals do. In this case, the old advice "feed a cold, starve a fever" is wrong! If you have a fever, feed it if you can—your body will appreciate the extra calories to fight off the illness.

Desert iguana fever is just one of the exciting stories you can find in evolutionary medicine. Evolutionary medicine is a growing new field that uses the principles of evolution to understand health and disease. Evolutionary medicine has been used to understand infectious diseases of all kinds: evolutionary medicine studies on malaria, cholera, and HIV have reframed how we understand these infections.

The inappropriately named "lifestyle" illnesses, too, can be studied through the lens of evolution. We have revolutionized how we think about cardiovascular disease and diabetes because of evolutionary medicine. Evolutionary medicine often uses the concept of "mismatch" to explain chronic disease. Mismatch is the idea that our bodies evolved under circumstances that were vastly different than those now found in complex, modern society. Because biological evolution is slower than social change, our bodies may not always function optimally in the highly industrialized, sedentary, high calorie, unequal societies we have created for ourselves.

One example of mismatch in evolutionary medicine is Type II (or insulin-resistant) diabetes. When the body cannot use insulin effectively to bring glucose into the body's cells and the glucose stays in the bloodstream, Type II diabetes results. Diet, particularly excessive intake of simple sugars and high fat foods, contributes to obesity, which is a risk factor for Type II diabetes. Humans probably evolved a preference for sweet and fat foods because these foods supply easily accessible, calorically dense nutrition. However, these nutrients are not commonly found among hunter-gatherer foods, meaning that although humans evolved a preference for them, they were likely not overeating them in the hunter-gathering subsistence systems that humans evolved in. In many societies today, industrialized food systems have favored high fat and sweet foods for a low cost, hoping to make money on our evolved tastes. Because human physiology did not evolve for this near-unlimited supply of foods, it accommodates excessive calories by storing the fat and reducing the ability of insulin to respond to glucose, favoring the development of Type II diabetes. This mismatch is compounded by a mismatch in physical activity, with modern work and car-focused cities combining to lower activity for broad swaths of the global population. Illnesses like Type II diabetes are considered classic examples of evolved physiology being mismatched with industrialized environments.

Finally, evolutionary medicine has been used to study women's health: topics ranging from obstructed labor and the use of birth attendants, postpartum hemorrhaging, and cesarian section have been explored from an evolutionary point of view, usually using a mismatch framework. For example, women who have the means to regulate their fertility with barrier methods (condoms) will probably have fewer children across the lifespan than women without them. There is a surprising health risk involved—these fertility-regulating women are at a higher risk of ovarian cancer. Repeated pregnancies and breastfeeding across the reproductive lifespan suppress ovulation. When a woman ovulates more frequently, the ovary faces repeated wound-repair events as the egg erupts from its follicle each month. Repairing this damage risks changes to cells that may become cancerous (Trevathan and Rosenberg 2020). Interestingly, women who use hormonal birth control, which stops ovulation from occurring, have a lower risk of ovarian cancer. Other topics, such as the use of midwives and birth attendants see a continuity between our evolved need for help during birth, and how that help manifests in current society (Trevathan and Rosenberg 2020). Examining the relationship between our evolved bodies and our current lived contexts—infectious, environmental, behavioral, social, cultural, and more—makes evolutionary medicine a powerful tool to help understand the whys of getting sick, and not just the hows.

The mismatch perspective in evolutionary medicine is vital for examining women's iron deficiency. Women's evolved bodies are mismatched to certain modern human contexts, particularly social contexts, meaning that illnesses that result from iron deficiency, like anemia, do not affect all people equally. We'll be examining the social causes of iron deficiency anemia in this book, keeping in mind that evolutionary mismatch underlies these social inequalities. Our evolved iron physiologies, while flexible, cannot operate effectively in a society that cannot properly support women's iron nutrition.

Life is a process

A simplified version of the Modern Synthesis—that alleles are passed from generation to generation—gives short shrift to what happens during the lifespan of the organism. It barely considers how organisms change during their lifetimes in response to the environment, including their growth, development, and reproduction. It turns out these processes, too, have evolved, and these evolved systems are incredibly important to the survival and fitness of organisms. Minor differences in growth rates, spacing of births, numbers of offspring, and timing of offspring birth can impact how much an organism reproduces across their lifespan. These tweaks can have evolutionary implications for the survival and fitness of an organism. If these parameters are tweaked too much, they can potentially impact health.

Life history theory is an offshoot of evolutionary theory, one that attempts to explain why different organisms have different patterns of biological life events. Why do mice grow quickly, reproduce early, often, and abundantly, and die quickly? On the other hand, why do humans and other primates grow slowly, give birth to one infant at a time, and have long lifespans? Life history theory explains that each lineage evolved unique biological patterns across their lifespan to optimize their fitness.

Life history theory starts with the idea that organisms have a certain amount of energy available to them as they live their lives. These life histories were shaped by natural selection to best allocate this energy across the lifespan in way that best balances survival and reproduction. Mice use their energy quickly, by reproducing quickly and having many offspring. The downside to this is that mouse lifespans are short and infant survival is low. Humans and other primates, however, use their energy more slowly across the lifespan and have slower growth rates, lower infant mortality, and longer lifespans. In between these extremes, there are a multitude of possible life history patterns among plants and animals.

According to life history theory, each of these life histories evolved to optimize evolutionary fitness. This theory is not only used to explain differences between species, however. It is also used to explain variation within a species too. In humans, life history theory has been used to explain why girls have their first periods at different ages, why women's births are spaced at different intervals, why there are differences in family size, and even why we grow old and die. We humans have similar, but not the same, biological events organized over our lifespans. Life history theory provides a way of explaining these biological differences between people from an evolutionary perspective.

You are probably thinking that the best life history is one that maximizes everything: bigger body, early reproduction, multiple offspring, low mortality, and an infinite lifespan. Life history theory puts limits on the energy that is available to an organism. Sometimes an individual has limited energy sources available in the environment. Sometimes the individual's physiology can only process a certain amount of energy for use. Yet other times, extra energy is stored as fat. Energy limits place real restrictions on organism function. When energy is used for one thing, it cannot be used for something else. This means that organisms must trade off energy from one function to another.

A central part of life history theory is the expenditure of energy across the lifespan. Biological organisms have metabolisms that take in energy from the environment and expend it on bodily functions across their lifespan. These functions are growth, reproduction, and maintenance of bodily systems critical to life. Organisms' physiological systems allocate[1] this energy to these three functions across the lifespan in a way that attempts to maximize lifetime fitness. However, since energy is limited, organisms' physiologies must make trade-offs. If they become infected, for example, the immune system mounts a response, leaving less energy to be devoted to growth. Reproduction and growth are both energetically costly, which is why they rarely occur at the same time in humans. Humans who do reproduce before growing, such as teenagers who become pregnant, are often shorter and have a greater chance of having high-risk pregnancies, indicating that there is not enough energy to devote to both finish growing and devote full energy to developing a full-sized, full-term fetus.

[1] When discussing life history trade-offs, scholars will often use language that suggests that the organism has agency over life history trade-offs: "allocates" is an example of this. When this language is used, it is not intended to indicate that organisms can choose the course of their life history; rather, the organism's physiology is altered in response to environmental context so that these trade-offs occur. No conscious thought or decision-making is meant to be implied by this choice of language.

While enacting trade-offs between reproduction, growth, and somatic maintenance, the body may alter physiological systems to achieve these goals. Most of these systems have evolved to be flexible in the face of environmental perturbations. Physiology tends to try to maintain homeostasis, or a steady internal state. Body temperature, fluid balance, breath and heart rate, the release of hormones, the circulation of immune cells, and the amount of iron in the body all attempt to stay within certain physiological limits. Sometimes, the environment shifts in such a way that homeostasis may be altered. For example, a higher rate of breathing at high altitude is a way of compensating for the less pressurized oxygen in the air. Life history trade-offs can sometimes also alter homeostasis. When the body encounters an infection, energy becomes devoted to activating the immune system and fighting the infection, while less energy may be devoted to, say, bone growth. This change in physiology in response to environmental pressures is referred to as a functional adaptation. In contrast to a genetic adaptation, which is a permanent part of the DNA of the organism, a functional adaptation tends to occur during the lifetime of the organism and is not encoded in DNA. This means it is not passed from parent to offspring in the Mendelian sense.

Although functional adaptations are frequently written about as positive forces in the functioning of an organism, they are not always harm free. Both life history trade-offs and functional adaptations can come with costs. This means that physiological adjustments, and indeed life history trade-offs, are not always optimal. Sometimes environmental pressures are too extreme, or life history trade-offs are too large to achieve. This phenomenon has sometimes been referred to as an accommodation—while the organism is still alive, it has made non-optimal adjustments that negatively affect its life course. An example of this would be extreme growth restriction in the face of infection or lack of food. While the body has accommodated the energy restriction by not growing as well, a smaller body size has significant risks for future reproductive success, ability to respond to future environmental perturbations, and later health. As I will demonstrate in this book, iron deficiency has similar consequences for women across their lifespans, too.

This is why failure to address public health concerns like food insecurity, sanitation, vaccination, supplementation, maternal health, and more, is a grave injustice—while the human body can often survive these insults, the results are subpar and sometimes even deadly. When applying evolutionary theories like life history theory to human conditions, it is vital to remember that just because a biological phenomenon *can* happen, that doesn't mean it *should*. This guiding principle will be used throughout the book—maternal iron deficiency is not an inevitable consequence of human functional adaptation. Social, economic, and political forces play just as much, if not more, of a role in placing our evolved bodies in situations that make them vulnerable to physiological accommodation.

There are three key ideas to take away from this introduction to human evolution. The first is to remember that humans (like all living organisms) are variable. This variation is shaped at multiple levels of biological organization—genetic, epigenetic, anatomical, physiological, behavioral, and more. Human variation is normal, is a function of complex human environments, and can be explained using evolutionary theory. This is, essentially, also the second key idea: that the environment is central to understanding how human variation may be shaped into predictable patterns. Evolutionary theories can help us understand biological patterns. However, social theories are critical to truly understand the human environment. Frequently, human biologists like myself turn toward that catch-all term, "the environment" to help explain human contexts. But we are not merely affected by the temperature, altitude, and sun; we are also deeply affected by the social world around us, including social structures that shape complex society. Human evolution can best be understood in the context of human social forces and the social sciences that study them. The third and final key idea is the understanding that not all human variation is inevitable, especially when that variation results in suffering. When human variation is the result of social inequalities, conflict, and disaster, we cannot bury our heads in evolutionary theory and accept adaptation as inevitable. Instead, we have an ethical obligation as scientists to serve the public, including remembering that our evolutionary insights could be used to improve the well-being of the public. This applies to women's iron deficiency, too.

Human context: History, society, and culture

Human social contexts matter to human evolution. Because humans are so social, human relationships, culture, and society are paramount to understanding how evolved bodies work—or don't work—in the context of health and disease. Complex societies—the kind that all humans are impacted by, even if they don't directly participate in one—are likely different in significant ways from the contexts of human evolution. For one thing, the advent of agriculture around 10,000 years ago allowed populations to become more sedentary and more densely clustered in settlements and changed human diets dramatically compared to the dispersed hunter-gatherer groups that likely formed the basis for most early human societies. Complex social structures, which are social institutions and patterned institutional relationships, have created a whole new set of environments for human existence. Complex societies are the types of environments that could be considered a mismatch—our bodies evolved in certain social and ecological settings and now they live in environments (of our own creation!) that are vastly different from the environments of human evolution. If you really want to synthesize evolved mismatch in human health and disease, you should understand the social sciences as much as the biological sciences, a difficult trick to pull off.

Believe it or not, evolutionary sciences and social sciences share many of the same intellectual roots. Both disciplines were grounded in hierarchical understandings of their respective phenomenon in 18th and 19th-century European-derived intellectual traditions. In biology, the systematist Carolus Linnaeus organized species hierarchically; while in social sciences, Lewis Henry Morgan organized societies hierarchically based on technology, family system, and material culture. In both disciplines, there was an underlying theme of progression; that is, moving "higher" in the hierarchy represented progress. In the case of the social sciences, Morgan considered European societies to be the "highest" form of society, while hunter-gather groups ranked the lowest. From these rankings, it was easy to conclude that the people within these societies were superior or inferior, just due to the type of society they belonged to. The belief that there were distinct races of humans, some that were superior to others along physical, intellectual, and moral dimensions, is now referred to as "race science." While there were variations on this theme—Darwin, for example, did not rank species hierarchically but did consider races of humankind to be hierarchically organized, separated via sexual selection in which in-group traits were preferred—this was the general trend of scientific thought during the 19th and early 20th century. It affected nearly every academic discipline. This includes the growing professionalization of medical practitioners, at least in the United States.

Race science was not a minority position among scientists in the 18th, 19th, and early 20th century. Instead, it was a recapitulation of the dominant European social structure of the time. Ideas of hierarchy, the superiority of one race over another, and progress were all ways to scientifically justify what was happening across the globe: colonialism, enslavement, and eugenics. That the science of the time, performed by male European-descended scholars, held up the systems that benefited them could hardly be considered a coincidence. This is a lesson for all scientists—science is a social endeavor, subject to major social movements of the time, the desires of institutions, and beliefs that serve the purpose of the powerful within society. It may not surprise you, then, when I note that strands of racist and colonialist thinking continue to crop up in later intellectual movements and social structures. The names might change but the practices and belief systems have core consistencies linked by historical threads.

In the social sciences, it is normal for new theories and ways of understanding to supplant old ways of thinking, sometimes in a reactionary fashion. In anthropology, race science was challenged by Franz Boas, in the early 20th century, whose groundbreaking work on immigrants showed that cranial shape—widely used as a marker of racial identity—was predicted by environment rather than being inherited from parents (Boas 1912). He also proposed a concept of culture that is still being used in some anthropological theories today: that culture is a body of habitual behavior, passed on via tradition and rationalized by a system of meaning. Boas's students elaborated on this by integrating innovative ideas from the field of psychology. Students like Ruth Benedict and Margaret Mead proposed that culture emerged from cognition, meaning that culture was an aspect of psychology. In contrast to earlier social scientists, he also developed the concept of cultural relativism—that societies cannot be placed above or below one another, and their cultures must be considered on their own terms rather than in comparison to others. Through his work and through the heavy mentorship of students, Boas pioneered anthropology as practiced in the United States (excluding physical anthropology, which will be discussed more later).

In addition to Boas's contributions to defining and understanding culture, an increasing focus on structure was also occurring in the social sciences. Structure is the arrangements of institutions within society. These structures both emerge from, and shape the behavior of, individuals within society. Karl Marx was one of the earliest theorists of social structure, connecting religious and political life to the "mode of production," which are the materials and methods used to create subsistence items. According to Marx and his intellectual partner Friedrich Engels, those that control the mode of production shape how an economy is structured, which in turn profoundly shapes the overarching social structures of society, particularly economic class. Subsequent sociologists such as Herbert Spencer, Émile Durkheim, and Max Weber also formulated theories on how social structures developed, functioned, and associated with each other within society. In early to mid-20th century anthropology, structuralist-functionalists such as Alfred Radcliffe-Brown saw societies as a set of complex parts whose function was to promote stability; while structural anthropologists such as Claude Levi-Strauss believed that cultures had deep structures that were embedded in the cognitive architecture of the human mind. Pierre Bourdieu extended this idea with the concept of habitus. Habitus is a set of internal structures or habits that are common to a social group, both shaped by external structures and contributing to them—this theory of practice continues to be a highly influential mode of linking social structure with individual agency. Early to mid-20th century theories, however, tended to lack full explanations of how social structures were designed to place some groups of individuals in society above others, such as on the basis of race or gender. A notable exception was the sociologist W.E.B. Du Bois, who discussed the concept of the racism of institutions, and how it impacted the health of Black populations in the United States as early as 1906 (Du Bois 1906).

Anthropologists began to use an anti-racist perspective during the 20th century. Change was slower in physical anthropology, where racist typology persisted. Earnest Hooton at Harvard, in particular, trained an entire generation of physical anthropologists on racial categorization. However, some researchers did resist the scientific racism of physical anthropology. W. Montague Cobb spent much of his career disproving scientific racism, although his position at Howard University's medical school meant he did not train anthropologists. Mid-century anthropologists such as Frank Livingstone and Ashley Montagu helped reimagine and popularize the distribution of human variation across time and space, replacing racial categories with clinal variation in traits and genes across geography.

In the evolutionary sciences, Stephen Jay Gould and Richard Lewontin challenged multiple paradigms that were emmeshed in typological thinking. In addition to tackling scientific racism, they also challenged the primacy of adaptationist thinking. The adaptationist paradigm is the idea that every feature seen in a living organism evolved via natural selection and is adapted to its current environment. In this paradigm, a "story" of an adaptation could be created to explain features without demonstrating differential survival and fitness. Gould and Lewontin proposed that there are many features on an organism that do not influence evolutionary survival and fitness but instead may have arisen due to other biological characteristics of an organism, such as anatomical or developmental constraints. Interestingly, this is an argument that places primacy on structure for shaping the biological attributes of species, an interesting parallel to the study of structure in social science.

They also made clear that anti-racism and anti-adaptationism are more alike than they seem. For example, Lewontin argued against a specific strain of behavior genetics, one that proposed that there were genes for behaviors and that these genes evolved via natural selection; worse, this had been picked up by race scientists who posited that these behavioral genes were characteristics of racial groups that had distinct evolutionary histories. Lewontin instead demonstrated the structure of genetic variation in populations, showing that between-group variation accounted for little of the total human genetic variation, and within group variation accounted for a lot of it. This means that it is impossible to distinguish "races" using any one gene, including any genes that may be responsible for behavior.[2] He concluded that evolution could not be used as an explanation for racial differences. While Gould and Lewontin's ideas were attacked by other scientists based on their social ideology, their ideas remain scientifically sound. Today, an anti-racist approach to human biological variation is mainstream in scientific circles.

However, sometimes the threads of old theories remain, invisible to the White majority but damaging to minority groups. Scientific racism is pernicious, and scientists within majority groups have never been that adept at reflecting on the practices of science and how they reify and reinforce prevailing social structures. The practice of trying to determine "ancestry" categories[3] from a skeleton has persisted in skeletal biology, for example, even though these categories poorly capture existing variation in populations, recapitulate racist understandings of humankind, and likely harm groups that have been racialized (DiGangi and Bethard 2021). And despite major advances in our understanding of the human body, health disparities persist around the world. Iron deficiency and anemia are also distributed unequally across gender, race, and class lines, strongly signaling that it is just as social as it is biological. In many places these inequalities reflect the global and local effects of colonization efforts by mostly European countries over centuries. Even though these inequalities have been very well documented by scholars, they are firmly entrenched, affecting the well-being of humankind.

Since the mid-20th century, a proliferation of diverse social theories continues to document human culture and society. While this chapter cannot possibly review the entirety of social theory in anthropology, I will highlight a few to give you a sense of the breadth of anthropological thought that can be used to understand human biology. In addition to the anti-racist movements just mentioned that gained steam in the middle of the 20th century (including critical race theory), feminist theories began to gain ground in anthropology in the 1970s. These works began the process of reinterpreting many formerly strongly held scientific findings and helped change the nature of scientific practice to include women. Other feminist theories, such as Black Feminist Thought, focused uniquely on the experiences of Black women. French

[2] Note that based on our current understanding of genes and behaviors, there are no single gene-single behavior associations. Rather, behaviors are a product of the complex of genes that determine the human brain, development, and environment—physical, social, and more.

[3] Race, in this case, was renamed "ancestry," possibly to shield practitioners and/or outsiders from the true nature of the practice.

theorists, such as Michel Foucault and Pierre Bourdieu, theorized on the nature of power—how it shapes the nature of social relationships, institutions, and individuals within society. Another theoretical thread concerns globalization, nation-states and the movement of people, capital, technology, ideas, and media across the globe, led by theorists such as Arjun Appadurai. Appadurai and Faye Harrison have also interrogated the production of theory in "prestige zones" within universities of the Global North, emphasizing that decolonizing anthropology would involve emphasis on theory production by scholars in the Global South (Harrison 2016). Recent trends in environmental anthropology emphasize ecological connection, climate change, and using the "point of view" of different species using actor-network theory to shed light on the human condition (Tsing 2015). In medical anthropology, structural violence, or the harm done to individuals by social structures when these structures fail to meet their needs, remains a highly useful theory for understanding how inequalities in health emerge between groups. Theories of embodiment, too, suggest how the social world "gets under the skin" to impact health, centering the social world and not genetics as the cause of biological differences between groups.

This is a lot of social theory! The theory that a social scientist chooses to use is highly dependent on the research question, the nature of the social phenomenon, and the lens through which they think the best interpretation will emerge. Biocultural anthropologists attempt to synthesize social theories, as best they can, with the social and biological phenomena they observe. Biocultural research has a long history in anthropology, dating to at least the Boasian school in the United States. It has been conceptualized as a bridge between the biological sciences and the social sciences, bolstered by the holistic aspirations of four-field American anthropology. Traditionally, biocultural anthropology tended to be centered within biological anthropology, with practitioners attempting a synthesis between biology, culture (loosely defined), and the environment. Under this framework, human adaptability to the environment was a central part of the biocultural endeavor.

In 1980, Alan Goodman and Thomas Leatherman developed what they called the New Biocultural Synthesis, which integrated Marxist political economy—an analysis of politics, economy, and society that centers the production of capital, access to resources, and class—with human adaptability and biological outcomes (Goodman and Leatherman 1998). This has been a highly influential approach within anthropological human biology. Another biocultural theory has been developed from Boasian concepts of culture, situating culture as emerging from the psychological and cognitive frameworks in human minds. This theoretical orientation is referred to as cultural consonance, and relies on the concept of cultural consensus, or collective shared knowledge about society. Cultural consensus is then used to elucidate how closely an individual thinks and behaves relative to a group's shared cognitive cultural model. The cultural consonance of individuals to their group's shared beliefs can be related to biological outcomes (Dressler 2020).

Critical Biocultural Anthropology is a more recent iteration of the New Biocultural Synthesis, one that seeks to integrate medical anthropology and the extended evolutionary synthesis more explicitly into the original framework (Leatherman and Goodman 2020). It is cognizant of the need to understand the interplay between human and environment, understanding that humans shape environments as much as environments shape humans. It attempts to integrate history and structural forces with human culture to understand the forces that shape local biologies, which are the process and product of intertwined social and biological experiences in a given time, place, and people. It also seeks integration with the extended evolutionary synthesis, attempting to incorporate soft heredity in multiple forms (non-DNA inheritance including epigenetics and other developmentally induced forms of biological variation) into human evolutionary biology. And it retains close ties with medical anthropology, particularly incorporating the concept of embodiment—the mechanism through which social forces become biology. The concept of "pathways of embodiment," taken from eco-social public health theory (Krieger 2001), attempts to distinguish how the social world "gets under the skin" and becomes biology. Presumably, other social theories can also be integrated into the framework as they apply to a given context. The ultimate goal of Critical Biocultural Anthropology is to capture, as holistically as possible, the human biological experience.

While Critical Biocultural Anthropology is a call to address the complex realities of the biological body in the social world, its downside is the difficultly of formulating a research project that can reliably grasp all aspects of the phenomenon under study. This is especially felt in the relative lack of work that successfully shows a "pathway of embodiment" between human bodies and human worlds. While biocultural theory is compelling, anthropology continues to struggle to capture complex human worlds. This problem is multifold, with primarily biologically oriented researchers failing to capture the complexity of the social milieu, and primarily sociocultural-oriented researchers conceptualizing the human body as a "black box," with no need to peek "under the skin" to see the human variation underneath. A major issue is connecting structural inequalities to embodiment in a meaningful way, since structural inequalities are so vast, and the process of embodiment is occurring at the cellular or physiological level This is exacerbated by the fact that scientific fields that are technologically better equipped to address these pathways, like biomedicine, still sometimes default to deterministic genetic and racial explanations for human biological variation. The solution requires interdisciplinary thinking and slow science; an approach that is not incentivized in fast-paced publish-or-perish academic contexts. But it is exactly what is needed to grasp the full extent of complex inequalities like the current state of women's iron status in the United States and across the globe.

The tesserae of iron

The ruins of the Amazigh city Volubilis stand north of the city of Meknes in Morocco. Nestled between green hills, Volubilis hosts impressive remains of Roman arches, public buildings, streets, aqueducts, and homes. Scattered throughout the ruins are beautiful mosaic floors, remnants of the wealthier homes in the city. The mosaics are beautiful: a man riding a horse backward, Diana startled at her bath by Actaeon, dolphins, the labors of Hercules, and more. When we look more closely, we can see that each mosaic has been pieced together with thousands of small, colorful tiles (Figure 1.1).

In mosaic-craft, these small fragments are called tesserae. On their own, each tessera is a colorful bit of glass, stone, or ceramic; sometimes carefully shaped to fit other tesserae, sometimes formed as a cube, sometimes rough. Alone, each represents hard work: all are carefully shaped, selected for texture, and painted. Together, they are assembled to form a complete picture, affixed with mortar and grout, polished, and put on display. This artform stands the test of time at Volubilis, drawing on small, well-crafted tesserae to resist crumbling within their mortar.

Critical biocultural anthropology is mosaic-craft: when phenomena are too complex to be captured by one study, it must rely on smaller solid pieces, brought together by theory and interdisciplinary connections. The result is small fragments of research delicately shaped into useful bits of anthropology. These fragments, or tesserae, are small glimpses into biocultural phenomenon. But tesserae never stand alone. When tesserae are bound together and filled in with grout, the resulting mosaic reveals a much bigger, multidimensional picture than any one fragment alone. Biocultural theory, with its lofty goals, must embrace this approach. Most studies can never be a complete mosaic by themselves—each tessera expends significant time and money. Researchers, recognizing the balance of completeness versus cost, tend to focus on achievable research problems. Some tesserae may seek to connect large concepts like political economy and biology; while others are focused on experiences,

Figure 1.1. Mosaic floor depicting the labors of Hercules in the ruins of the Roman city of Volubilis in Morocco.
Source: Photo taken by the author.

cognition, behaviors, or physiology among smaller groups of affected individuals. Some may drill even deeper into biological or sociocultural understanding with advanced qualitative or laboratory techniques, while others use mixed methods to grasp parts of both at once. The size or scope of each tessera is less important than the understanding that each brings to the greater whole. The goal of biocultural anthropology should be to bind and fill these tesserae, these useful fragments, into a fully realized picture.

This is the lofty goal of this book. What we know about women and iron are small pieces: single-minded understandings of supplementation and biomedical management, iron physiology, the iron demands of pregnancy, racial and socioeconomic disparities in iron status, a global iron deficiency crisis. Slowly, we are adding more tesserae: how evolution shapes women's iron status, the gut as a nexus between the outer world and embodied inner world, historical conceptualizations of iron status and how they became racialized and gendered, the relationship between iron and the mind. The mosaic of this book is to weave them together in a biocultural whole, pulling pieces together to shape our biocultural understanding of iron deficiency and why it is so widespread.

Chapters 2–4 take an evolutionary approach to understanding women's iron status. In Chapter 2, I will discuss the evolution and physiology of iron in the human body to help form the basis for understanding iron biology in later chapters. Iron physiology walks a narrow tightrope between helping and harm, and the evolution of iron across the history of Life is delicate funambulism. Chapter 3 centers around evolutionary approaches to iron physiology during women's reproduction. Iron is used heavily during pregnancy, with consequences for women's life histories and ultimately, their survival. Chapter 4 explores why women have lower iron levels than men. Biomedicine blames menstrual periods—but is this true? I will unpack the framing of menstrual periods as pathological using an evolutionary and biocultural approach.

Chapters 5–7 use social theory to understand how iron deficiency has been constructed as a disease. Underlying this approach is the evolutionary medicine concept that there is a fundamental mismatch between social factors and what women's bodies evolved to do. Chapter 5 is centered on the embodiment of iron status. I will talk about the gendered ways that iron gets into the body, how it gets out, and the role of the gut (and its microbiome!) in the dynamics of iron embodiment. Chapter 6 delves into the history of how iron deficiency was conceptualized as an illness, with a special focus on race, racism, and the United States. Unsurprisingly, concepts of iron status and illness mirrored colonialist and racist thinking, with tendrils that continue to extend into present day biomedicine. Finally, Chapter 7 frames iron as part of a global nutrientscape that is distributed unequally both within and between bodies, including the surprising links between mental health and iron status in women. This chapter asks: What social structures have shaped global patterns of the hidden hunger of iron deficiency? What subtle harms are happening to the world's women as they live with iron deficiency? Woven throughout each chapter are the reciprocal links between the biological and the social, the way evolution and social structure fit in one mosaic image.

Concluding thoughts

Iron deficiency is harmful to women. One-note solutions—like supplementation—have not solved this global crisis. Anthropological theory is a powerful toolkit to take apart the simple explanations and to rebuild a more complete mosaic of this crisis. Without it, iron deficiency will continue to reoccur to the detriment of the Earth's women. A true solution lays bare the world's wretched social inequalities and the evolved limits of the human body; any other approach is merely a small bandage on the hemorrhage of suffering wrought by human greed. The violence meted out by our constructed social world on women's iron physiology is invisible; a public health priority but one that is poorly understood and difficult to act on. This violence is deadly to the 591,000 + women who die each year of iron deficiency while they gestate, give birth to, and raise their babies.

This book is ambitious—as ambitious as constructing a Volubilis mosaic. Throughout the book, I've attempted a true interdisciplinary approach, from biochemical evolution to history and humanities. Each piece is small but the whole glimmers with light. Let it reflect the need and the hope of women suffering from iron deficiency; may it find a way to improve their lot.

2
So Metal
How Iron Supports Life

The first tesserae of this book are pieces of the lustrous silver-gray iron, outlining the image and weaving their way throughout the design. The iron gives our mosaic a curiously muted shine, as if there is more to it than meets the eye. Life depends on this iron. But it doesn't exist as a solid element in biological organisms, nor does it stay that way if exposed to water or air. Iron is chemically reactive: it easily picks up electrons from other elements and transfers them, making iron extremely useful to the metabolisms of cells. On the other hand, this reactivity is risky. Unchecked chemical reactions can stress and damage biological systems. Iron's dangerous charm, the desideratum tied with menace, is the forge through which biological organisms evolved. The story of iron evolution is one of mitigation, maximizing the benefits of iron while minimizing its harms. This story begins billions of years ago on Earth.

Fe

Iron (chemical symbol: Fe) is abundant throughout the universe, produced in the last gasp of dying stars. When supernovas collapse, the energy from fusion creates iron and scatters it throughout the cosmos, until it eventually becomes part of rocky planets like Earth. Iron is the most abundant element by mass within Earth. It does not lie there inert, but instead reacts continuously with the most abundant element by number: oxygen. Given iron's reactivity, iron is also vital to the function of all living organisms, including humans. The relationship between iron and biology is one of the most intricate, long-lived stories of evolution.

Iron is rarely found in nature as a pure metal. Its atomic configuration is too volatile to remain stable. Instead, it easily moves electrons to other atoms. Electrons, the tiny negatively charged particles that circle the positively charged atomic nucleus, can leave one atom and join others, forming chemical compounds. Iron's particular electron configuration means that it can both

Thicker Than Water. Elizabeth M. Miller, Oxford University Press. © Oxford University Press 2023.
DOI: 10.1093/oso/9780197665718.003.0002

gain and lose electrons easily. The loss of electrons, called oxidation, is more easily achieved by iron than reduction, which is the gain of electrons. Thus, it is common in nature for iron to lose two electrons, forming an ion with a positive charge. This ion is referred to as Fe^{2+} and is commonly called "ferrous" when it is in a compound. For example, rust is a compound consisting of Fe^{2+} and oxygen, forming the compound ferrous oxide or iron (II) oxide (FeO). When iron loses three electrons, it is referred to as "ferric," and its reaction with oxygen is ferric oxide or iron (III) oxide, with the chemical compound written as Fe_2O_3.

Iron is sensitive to the environment, and differences in temperature, pH, and the availability of other elements can affect how iron oxidizes. During Earth's first 2.5 billion years, iron compounds were formed as ferrous iron or Fe^{2+} lost two electrons and formed compounds, like pyrite (FeS_2) formed with available sulfur (Wächtershäuser 1988b). This reaction was facilitated by the volcanic nature of earth, as iron and sulfur react when they are heated. Iron did not form compounds with oxygen, because for the first 2 billion years or so, there was almost no oxygen on Earth. This seems impossible to comprehend for living, breathing, people. How did Life manage to emerge from a majority-iron planet without any oxygen?

The emergence of Life

The evolution of Life from non-living compounds—or abiogenesis—is shrouded in mystery. Abiogenesis theories hypothesize a jumble of compounds emerging from mixtures that contain carbon, hydrogen, and nitrogen, a concept referred to as a hot "primordial soup." The creation of these compounds was the first step in a complex process that led to the formation of the cell, which is the smallest organizational unit of life.

Certain properties are characteristic of all cells: being self-enclosed within a fatty (lipid) membrane, being able to self-replicate, and having a distinct set of successive chemical reactions that form the basis of cellular metabolism. Most theories about the origins of life center on the process of self-replication, putting RNA and its replication first in the evolution of Life (Robertson and Joyce 2012). However, some theories focus instead on the evolution of metabolism—the chemical reactions that put a cell into motion. Iron may have been part of the very first metabolisms on Earth. One hypothesis, the iron-sulfur world hypothesis, posits that early life emerged via the support of iron sulfide minerals. First proposed by Günter Wächtershäuser in 1988, this theory is in line with the prevailing view that life originated in hot, hydrothermal volcanic vents on the ocean floor (Wächtershäuser 1988a). These vents would have been made of iron (II) sulfide (FeS). The hot FeS assisted early metabolic reactions by accepting extra electrons and therefore eventually evolved to be part of cellular metabolisms. FeS is still used in the metabolism of many prokaryotic cells in the form of iron-sulfide proteins called ferredoxins (Mutter et al. 2019). In fact, several human proteins contain ferredoxins, remnants of our ancient common ancestor with prokaryotes. While "metabolism first" and "replication first" camps for the origin of Life are often in disagreement, the early integration of iron into protein—found in such disparate evolutionary lineages as bacteria and humans—shows that iron in the environment intertwined with Life quite early in evolution (Wächtershäuser 1992; Williams 1981).

Early cyanobacteria began producing oxygen about 2.4 billion years ago in what is referred to as the first Great Oxygenation Event. Much of this oxygen reacted with Fe^{3+}, creating large deposits of ferric oxide, commonly known as rust (Holland 2006). This "mass rusting" reduced bioavailable iron in the environment and was catastrophic for the prokaryotes who relied on it. However, oxygen-adapted species, using aerobic respiration, were more efficient at yielding energy from glucose (Sheftel, Mason, and Ponka 2012). This event coincided with the appearance of eukaryotic cells, which contain DNA within a nucleus. As oxygen in the atmosphere increased, Life evolved in complexity, with oxygen-associated compounds increasingly becoming part of the metabolisms of multicellular organisms. Iron was the center of it all: easily creating compounds with oxygen, it became part of the metabolism of fungi, plants, and animals. Iron intake became crucial to survival, and entire physiological systems evolved to take in and move iron throughout the bodies of multicellular organisms. Some physiological processes are ancient and shared by many biological lineages who share a common ancestor. Some are more recent, and therefore shared by fewer organisms. Iron solves problems, but it isn't harm-free. Evolution has had to balance the help with the harm, producing systems that keep iron's hefty chemical reactions from happening unchecked in the body.

Iron hurts

In humans and other animals, iron solves a major physiological problem. It prevents hypoxia, or the lack of oxygen in tissues. Hypoxia is caused by areas of low oxygen in the animal's environment or through deficiencies or infection that affect the body's oxygen absorption or transfer systems. Since human cells require oxygen, iron performs a vital service. However, unchecked iron causes tremendous harm.

The ease with which iron interacts with other elements is both useful and difficult for biological organisms. The reaction of O_2 and Fe^{2+} creates free radicals—an iron atom with a free electron that is not paired to another atom—which can cause damage to biological structures. Free radicals are unstable, constantly seeking to take an electron from another atom to make up for the missing half. Free radicals take electrons from anything they can, including protein, lipids, and even DNA. Free radicals can damage DNA base pairs or even break DNA strands. Free radicals can also set off free radical chain reactions among other atoms within cells. This type of damage is known as oxidative stress. This stress can be balanced by antioxidants, which are molecules that donate electrons easily to free radicals. Cells can overcome mild to modest effects of free radical reactions. However, without systems to control the levels of free radical–producing atoms such as iron, organisms are at risk for cell death. When cell death is widespread throughout tissues, organisms are at significant risk.

Because of the damage oxidative stress can cause, it's unsurprising that it is implicated in multiple chronic illnesses, including cardiovascular disease, Alzheimer's disease and other neurodegenerative diseases, cancer, and depression (Pizzino et al. 2017). These harms persist because iron is just that good at moving oxygen through the body. How did evolution help physiologies balance an atom that is so crucial to life but so dangerous to manage?

The answer to this can be found in early evolution. Bacteria and archaea have several metabolic pathways for managing iron, utilizing its ability to easily form compounds as part of the chemical reactions that sustain their metabolism. As Life evolved over time, natural selection shaped the pathways through which iron is utilized, embedding it within proteins, salts, and chelates, and keeping it from easily forming free radicals. These evolutionary novelties merely mitigate the danger, however, and organisms must constantly negotiate with the metal that embedded itself into biology. In humans, a complex web of proteins escort iron throughout the body to keep free iron from causing damage to delicate cells, preserving (as much as possible) the integrity of DNA.

The second major cause of biological damage due to iron is infection. Iron benefits both pathogens and hosts: bacteria need iron as much as humans do, and in the presence of free iron, bacteria often win the game of iron absorption. Some of the most harmful bacteria to humankind benefit directly from iron. Many bacteria in the phylum Proteobacteria, for example, are especially harmful, releasing proteins that can cause tissue damage and instigate inflammation. Unsurprisingly, many of these bacteria also have serious iron scavenging abilities.

There are multiple pathways bacteria can use to take in iron—such as finding ways to extract it from host iron proteins—but the most used path is the production of siderophores. Siderophores are proteins secreted by bacteria with a high affinity for iron, much higher than the host's iron proteins. Siderophores chelate iron, which means that they bind to metal ions. This has the effect of sequestering iron, keeping it out of reach of the host's physiology. Siderophores can also make the iron ions more functional to bacteria, reducing non-bioavailable Fe^{3+} to the highly bioavailable Fe^{2+}. Bacteria recognize the siderophore using specialized receptors on their surfaces and take up the iron, and it is released for use by the cell.

This process is efficient and helpful for bacteria who produce siderophores and use iron in their metabolisms. This is true of many members of Proteobacteria, whose populations have been shown to bloom in human microbiomes when excess iron is present (Chapter 5; Jaeggi et al. 2015). This is not necessarily a good thing for their human hosts, particularly when the Proteobacteria organism in question happens to be an infectious pathogen. Experiments have shown that the virulence of many pathogens, including those in the genera *Aeromonas*, *Clostridium*, *Corynebacterium*, *Escherichia*, *Klebsiella*, *Salmonella*, *Listeria*, *Mycobacterium*, *Neisseria*, *Pasteurella*, *Pseudomonas*, *Vibrio*, and *Yersinia*, are increased when iron is injected into infected organisms (Ratledge and Dover 2000). A higher virulence means that the bacteria are doing greater harm to the organism, either directly via the release of damaging compounds or indirectly by provoking an increased host inflammatory response.

The human body's response to infection can be just as damaging to the body as direct infection. Inflammation, an immediate immune response to infection, can be as damaging to the body as the bacteria. Normally, a short exposure to inflammation during an infection is beneficial to the body—any damage that is incurred is mitigated by the elimination of the pathogen. However, long-term exposure to inflammation is incredibly damaging, and so is an inflammatory response that is out of proportion to the infectious threat. Chronic inflammation is a continuation of short-term inflammation, including vasodilation and capillary permeability, increased blood flow, migration of immune cells, production of inflammatory cytokines, and tissue damage and repair. Diseases like stroke, chronic respiratory diseases, heart disorders, cancer, obesity, and diabetes are all chronic inflammatory illnesses, and they form the bulk of mortality due to chronic illnesses globally (Pahwa et al. 2021). Iron can make inflammation worse, meaning that its use by the body persists despite the damage it can cause.

Despite this, iron persists in the metabolisms of humans. This means that the benefits must outweigh the harms and that the body has evolved ways of mitigating the damage iron can cause. We will look at human iron physiology as an example of how biology negotiates the use of dangerous but useful substances. To do this, I will discuss how the body absorbs iron, moves iron throughout the body, uses iron, and gets rid of iron (if it can).

Iron in humans

Human physiology tightly regulates iron via homeostasis. Homeostasis is the tendency of physiological systems to maintain an equilibrium. In the case of iron, homeostasis is maintained when iron absorption matches iron loss. Biological anthropologists have been key investigators of homeostasis and its limits, often going to extreme human environments to understand human physiological adaptation. The investigation of iron homeostasis, however, has mostly been left to different fields—the exception being the investigation of red blood cell production in high altitude environments.

The recently discovered hepcidin molecule is a key player in our understanding of iron homeostasis (Erwin et al. 2008). It drives the absorption and regulation of iron in the body. Hepcidin was discovered in 2000, late in our understanding of iron metabolism. Initially described as LEAP-1, or Liver-Expressed Antimicrobial Protein-1, hepcidin was first isolated from human urine. Its first name came from the fact that it was expressed by the liver and demonstrated anti-bacterial activity. Most of our understanding of hepcidin function arises from studies of mice, where a notable experiment that knocked out the hepcidin gene in mice led to severe iron overload. Subsequent work identified the relationship of hepcidin to inflammation, anemia, iron stores, and hypoxia.

Before we fully discuss the action of hepcidin, we first should trace the path of iron into (and out of) the body (Sheftel, Mason, and Ponka 2012). The first port of entry of iron into the body is via food, where it is absorbed by enterocyte cells of the small intestine. Most dietary and supplemental iron does not get absorbed. Only about 10–25% is absorbed by the small intestine. Not all food is absorbed equally, either. Iron from animal sources, known as heme foods, is more bioavailable than iron from plant sources, or non-heme foods. Certain foods also help facilitate or inhibit iron absorption. Vitamin C, for example, enhances the absorption of iron if the two are ingested at the same time. Calcium and polyphenols, which are plant compounds with dietary effects in humans, can inhibit the absorption of iron. Taking your iron pill with tea, which contains polyphenols, and milk, which contains calcium, is a recipe for low iron absorption. Because most iron is not absorbed even in ideal digestive circumstances, it passes through the intestine and out of the body, unless the microbial inhabitants of the gut use it.

Intestinal enterocytes, which are cells that line the intestine, either store iron within the protein ferritin or release iron via the protein ferroportin. Ferroportin is the only structure that allows for the transport of iron from the inside of the cell to the outside of the cell. It is expressed only on certain cells of the body, including enterocyte cells of the small intestine, hepatocytes (liver cells), macrophages (immune cells), and adipocytes (lipid storage/fat cells). Hepcidin regulates ferroportin, and thus iron absorption, by binding to ferroportin and preventing the release of iron. Thus, hepcidin is the main regulator of iron homeostasis. When hepcidin is high, iron absorption is low (and vice versa). Hepcidin is sensitive to iron storage, meaning that when blood iron and iron stores are full enough, hepcidin is produced. This system helps prevent the body from experiencing iron overload and keeps the body within homeostasis.

Once iron is released from enterocytes via ferroportin, it does not run free throughout the body. It is bound to the protein transferrin and re-enters cells via transferrin receptors. Transferrin is only about 30% saturated with iron, meaning that any available iron that enters the body is quickly bound to transferrin. From there, iron is transferred and stored within ferritin in liver cells and immune cells. Transferrin and ferritin ensure that the body is not exposed to dangerous free iron while still maintaining safe and necessary iron stores. Animal bodies have other iron-transfer proteins in addition to transferrin that serve specific purposes. For example, lactoferrin is found in mammalian milk and helps transport iron safely to offspring, and ovoferritin is found in bird eggs and serves a similar purpose. But ultimately, serum transferrin is the dominant iron-transfer protein in human physiology.

The main metabolic function of iron is the transport of oxygen from the lungs to tissues. This is accomplished via the molecule hemoglobin (abbreviated Hb) found in red blood cells. Most iron in the body is in the form of hemoglobin. Hemoglobin is a ring-shaped organic molecule capable of holding one iron ion. Oxygen binds to the iron ion in the hemoglobin for transport to tissues. Iron is also a vital part of the production of adenosine triphosphate or ATP, which is the major energy-carrying molecule in cells. Between these two functions, iron plays a vital role in human energy metabolism. Other uses of iron in the body include creating myoglobin, which transports oxygen to muscles, aiding the proliferation of B- and T-cells in response to infection, brain functioning including myelination and neurotransmitter synthesis, the production of DNA metabolic enzymes, and growth and development (Beard 2001).

There is no physiological mechanism for ridding the body of iron. Once iron is in, it mostly stays in. Most iron that is lost is lost via dead red blood cells shed in stool and amounting to less than 1 mg a day. Most of the iron from red blood cell death, however, is recycled back into transferrin and ferritin to aid in the production of new red blood cells. Other causes of iron loss would be small (or large) amounts of bleeding in the intestinal or urogenital tract, via menstruation, or as part of the transfer of iron to the fetus during pregnancy. However, most iron does not leave the body once it is in. This is why iron homeostasis strictly regulates iron absorption. Allowing too much iron to enter the body would be a disaster—there would be no way to excrete it. When iron absorption is disrupted in either direction—either too much or too little—illness can be the result.

After all of this, it is no surprise that humans only absorb a fraction of dietary iron. From the way the scientific literature presents the relationship between hepcidin and iron homeostasis, it might seem like there is insignificant variation in the actions of hepcidin or the abilities of humans to absorb iron. If your iron is low, you absorb more and if it's high, you absorb less. However, this shields the fact that individuals vary in their baseline hepcidin levels. This is something that is not well understood at the population level, which is the level at which most biological anthropologists understand human biology. There is also variation in how iron is distributed throughout the body at the population level, which is also poorly understood except in one notable case: adaptation to high altitude (discussed more later in this chapter). There are also notable differences in iron status and broad population characteristics such as age and sex. A combination of genetic, environmental, and sex and lifespan differences contribute partially to the patterns we see at the population level for individuals that are considered in the physiologically "normal" range of iron status.

Far from being a general "one size fits all physiology" in all humans, iron homeostasis does vary at the genetic level, which results in individual variation in iron absorption. Genetic variants shape the ability to produce and use hepcidin. Some, like the C282Y mutation (the rs1800562 SNP) or the H63D mutation (the rs1799945 SNP) on the homeostatic iron regulator (HFE) gene are associated with lower hepcidin levels, meaning that more iron can be absorbed via the gut (Hollerer, Bachmann, and Muckenthaler 2017). Since iron cannot be easily metabolized out of the body, over time this can lead to iron overload in affected individuals, an issue I'll discuss in the next section. This is especially true of those with the C282Y mutation, which has a stronger effect on iron status than H63D. There are also genetic variants that cause too much hepcidin to be produced, making iron deficiency more common in those who have the variants. These variants are not as common as the mutations that lead to higher iron absorption, meaning that their true impact at the population level is unknown. Mutations in the transmembrane serine protease S6 (TMPRSS6) gene, for example, are associated with hereditary iron deficiency in some families. Although the exact mechanism is unknown, it is associated with high hepcidin (and low absorption of iron) and increased withholding of iron within macrophages. Mutations at bone morphogenic protein 2 (BMP2) and 6 (BMP6) alter the signaling pathways that produce hepcidin, meaning that these alleles may also be associated with iron deficiency (Canali et al. 2017). One study found that Saudi Arabian women homozygous for the rs235756 mutation at BMP2 were more likely to be iron deficient than those who were not (Al-Amer et al. 2021). Interestingly, this variant is reported to be common (about 0.33 allele frequency in one study), but the population-level implications for iron deficiency are not known. This is true for other alleles as well: alleles like the C282Y mutation, which have a more limited geographic distribution, are better characterized compared to alleles such as the H63D mutation that are more globally widespread. More work is needed to fully understand how iron absorption is modified by genetic architecture.

Iron distribution throughout the body is not well understood. We can deploy one of the core frameworks of anthropological human biology to try to understand it: functional adaptation of iron physiology to the environment. Unlike genetic adaptations, which are the result of natural selection making changes to allele frequencies over time in populations in response to the environment, functional adaptations are conceived as changes that are made to growth, reproduction, and physiology that allow the body to respond to the environment without changes in the underlying genetic architecture in the population. Human physiologies are quite adaptable, meaning they are frequently able to respond to extreme environmental conditions. One of the best-characterized extreme environments that influences iron physiology is high altitude. While few populations live more than 10,000 ft above sea level, those that do, or who travel there, have a physiological problem to solve: hypoxia. At high altitude, the air pressure is lower—that is, the molecules that make up air are farther apart—which means that the partial pressure of oxygen is lower too. Unfortunately, lungs do not increase in size enough to compensate for the lower partial pressure of oxygen, so physiology must compensate for lower oxygen intake somehow. Several functional adaptations "solve" the problem of hypoxia, including a genetic adaptation affecting Tibetan high-altitude natives, but I'll focus on the functional adaptations found in most who live long-term at altitude.

The biggest iron-related functional adaptation is erythropoiesis, or the formation of more red blood cells. This helps capture more oxygen in the lungs for transport to distant tissues. However, erythropoiesis requires iron to create the hemoglobin molecules in the red blood cells, which can be a problem for those residing long-term at high altitude. The chronic production of more red blood cells is known as erythrocytosis, and this can potentially lead to iron deficiency if dietary intake of iron is not met. In sea level natives who visit high altitude locations, iron stores are mobilized from ferritin to support the increase in red blood cell count. In fact, hypoxia is a well-known modifier of hepcidin levels, reducing them so that more dietary iron can be absorbed and more iron stores can be released from ferritin. However, iron deficiency due to erythropoiesis is not quickly replenished even with lowered hepcidin levels. The literature on iron homeostasis in high altitude natives is less clear. While long-term physiological adjustments to iron absorption and utilization in high altitude residents have been recognized, it is still not fully known how this occurs. Further, populations that are at risk for iron deficiency, such as pregnant women, are vulnerable at high altitude due to the slow replenishment of iron stores (Muckenthaler, Mairbäurl, and Gassmann 2020).

In addition to variation caused by iron-metabolizing genes and extreme environments, there are also clear average differences in iron biomarkers between men and women, as well as changes that occur in iron measurement across the lifespan. Notably, iron biomarker levels (Hb, ferritin, etc.) increase during childhood to puberty when sexual dimorphism emerges. In adulthood, men have much higher levels of both ferritin and hemoglobin than women. Women's hemoglobin and ferritin begin to rise post-menopause, but never quite reach the level of men's measures. It is commonly assumed that reproductive-aged women's lower iron biomarkers are due to the iron loss of pregnancy and the blood loss of menstruation, but this ignores a unique aspect of men's physiology—testosterone—that promotes erythropoiesis. Since women's iron is the core topic of this book, it will be discussed more in Chapters 3 and 4, where I will unpack the assumptions that have cast women's iron physiologies as pathologically low.

Evolutionary medicine of iron

The evolution of iron physiology emphasizes the danger that iron can cause to physiological systems. Iron is whisked from place to place in the body, constantly shrouded in protective proteins and cells that keep it from causing trouble. These systems are optimized when there is sufficient absorption of iron to maintain homeostasis. It might be unsurprising, then, that iron status generally tends toward "too little" iron, especially in women and children who have lower bodily iron levels. This means that these groups, especially, are at risk of being out of iron homeostasis when they are not able to absorb enough iron. However, there are cases where the body tends toward "too much" iron, and men are more affected by this than women or children. Can evolution say anything about iron homeostasis that is outside the range of optimal function?

When iron is not in homeostasis, people can become sick. Too much and too little iron can be a cause of illness. Too much iron can also contribute to infection. In response to infection, the body maintains homeostasis, in part, by sequestering iron away from pathogens. In this section, I'll discuss the evolutionary medicine of three different scenarios: too much iron, too little iron, and iron and infection. In all three cases, there are well-established biomedical "hows" that explain the mechanisms of illness. Luckily for us biological anthropologists, each case also demonstrates a different aspect of evolutionary medicine.

Case 1: Iron deficiency

I would call the case of too little iron an evolutionary mismatch—the iron coming in from the environment does not match the body's iron needs, possibly due to a current environment that does not "match" the iron environments generally found during human evolution. The immediate cause of the mismatch is either not enough dietary iron or something blocking the absorption of iron. The more distal causes of the mismatch are social conditions that prevent certain groups from being able to access sufficient iron, which I'll talk about in Chapters 5 and 7. For now, we'll talk about the physiology of low iron, its symptoms, and the consequences for human well-being.

The body is iron deficient when, for whatever reason, it does not have enough iron to meet its physiological needs. This means there is not enough iron in storage and not enough to meet the demands of hemoglobin production. When there is not enough hemoglobin, not enough red blood cells are produced. Anemia results when the body does not make enough red blood cells. Anemia has many potential causes, but iron deficiency is one of most common. Hemoglobin measurement is used to diagnose anemia, with hematocrit (the percent of the blood that contains red blood cells) being used to support the diagnosis. Below a certain threshold, a person is considered anemic. Note that the threshold differs between men and women—women have a hemoglobin threshold two grams per deciliter lower than men's, indicating that women have a lower baseline level of hemoglobin. This probably has you wondering what the reason for this is, but I'm going to ask you to hang on until Chapter 4. Hemoglobin, while correlated with the overall amount of iron in a person's system due to the sheer volume of red blood cells in the body, is not diagnostic for iron deficiency. You can have anemia without iron deficiency, and iron deficiency without anemia. To diagnose iron deficiency, you must look at other biomarkers of iron homeostasis.

By the time the body has too little iron to make red blood cells, body stores of iron are already quite low. Low levels of ferritin indicate that the body is iron deficient, demonstrating that there is little demand for iron storage. The amount of soluble transferrin receptors circulating in the bloodstream increases in an attempt to pick up more iron to transfer to cells. Finally, little to no hepcidin is produced, meaning that it does not bind to ferroportin which allows more iron to be absorbed via the small intestine. Ferritin, soluble transferrin receptor, and hepcidin are all good measures of iron deficiency in the body. However, ferritin and hepcidin can be modified by inflammation, meaning that not all measures of iron status can tell the whole story.

Iron deficiency, as you can imagine, is much less straightforward to diagnose than anemia. While iron deficiency is the most common cause of anemia, anemia can have other causes, such as Vitamin B-12 deficiency, and inflammation can alter iron status biomarkers in a way that masks true iron deficiency. Frequently, researchers will use combinations of biomarkers, such as soluble transferrin receptors and ferritin, in individual and population evaluations of iron deficiency. This is complex to interpret and goes beyond the routine lab testing requested by doctors. Despite these limitations, scientists believe that up to 2 billion people are iron deficient across the globe (Zimmermann and Hurrell 2007). Since iron is one of the most abundant minerals on Earth, how does this happen?

The simplest answer is that iron from the iron nutrientscape—my term for the distribution of iron across the environment—does not reliably make it into the human body. One simple explanation is a lack of iron-rich food availability, the types and amounts of food available to different people in different places. It may also be due to poor iron absorption, where diets that are rich in calcium or heavy in tea consumption inhibit the absorption of iron. Other causes could be hidden intestinal bleeding due to parasites, intestinal infections, microbiome disruptions, and other intestinal diseases. More complicated are the factors that contribute to the lack of iron-rich food and gastrointestinal disturbance—the cultural and structural portions of the nutrientscape that constrain who has access to what, including food, infection-free living space and water, and healthcare. These social inequalities exacerbate the iron status afforded by evolved iron physiology—while physiology throttles iron absorption, lack of available iron makes the problem worse. You could say that this is an evolved mismatch, because human iron physiology did not evolve in the context of global social structures that create poverty. While certainly illness and famine existed during human evolution, the circumstances that put people in poverty day after day and year after year are less likely to have occurred in low population density, highly mobile human ancestors. The iron nutrientscape that creates this evolved mismatch will be discussed in Chapter 7.

Case 2: Iron overload

In 1865 a French physician named Armand Trousseau treated an emaciated, feverish, 28-year-old man admitted to bed 3 in St. Agnes's ward (Trousseau 1870). The young man had been eating poorly since he lost his well-paying job two years earlier, but three weeks of increased hunger and thirst drove him to the hospital. He was drinking and excreting copious amounts of fluid that, upon testing, was found to be full of sugar. Diabetes was the clear diagnosis, but other features of his case stood out to Trousseau. He was struck by the bronzed appearance of the man's face and other discolorations of his skin and the patient's enlarged liver. Despite being treated with chalk and oxygen, the young man began to eat and drink less, eventually dying two weeks into his stay. Upon autopsy, Trousseau noted that his liver was twice the size it should be, had an unusual dense, firm texture, and was colored a greyish yellow. This was the first case study of the illness that was called first "bronze diabetes" and later hemochromatosis, after the bronze coloring that often accompanied it. Further work indicated that too much iron in the liver was, at least in part, responsible for the physical changes that accompanied hemochromatosis.

The case of too much iron is puzzling. Since the body evolved to make sure too much iron could not enter the body, how could iron overload exist? Iron overload is a recognized state, often (but not always) diagnosed as hemochromatosis, and it is recognized as harmful to the body in multiple ways. The main feature of iron overload is too much iron in what are called the parenchymal organs—the liver, kidneys, adrenal glands, pancreas, and spleen. Some people with iron overload, like Trousseau's patient, have so much iron that their skin becomes bronze, one of a common triad of symptoms, along with liver cirrhosis and Type II diabetes, that is still recognized today. Iron overload is known for being associated with too much blood glucose and is a recognized feature of certain metabolic illnesses.

Unsurprisingly, health issues like arthritis, cancer, and heart problems are also associated with too much iron. Iron overload can lead to impotence in men and amenorrhea in women, meaning that its effects on fertility appears to have a negative effect on evolutionary fitness. However, the group that is most afflicted by iron overload is older men. People with iron overload syndrome are more susceptible to bacterial infections, too, in part because the excess iron in their blood and tissues feeds infectious bacteria. This means that they are at greater risk of mortality, another sign that iron overload may not be favored by evolution. In the United States, the prevalence of iron overload is about 1 in every 200–500 people (MedlinePlus 2022). This is equal to almost a million people affected by iron overload in the United States alone. If this feature is so detrimental, why hasn't natural selection removed it from the population via its effects on reduced fertility and increased mortality?

This is another case that evolutionary medicine can explain, but this time it's not an evolutionary mismatch. Instead, most cases of iron overload can be explained by a single mutation. Hereditary hemochromatosis is a single mutation genetic disease on the homeostatic iron regulator (HFE) gene. In this illness, the body absorbs too much iron but has no way to get rid of it. This disease is an autosomal recessive disorder, meaning that affected people must have two copies of the recessive allele to have hemochromatosis. Carriers of the allele—those who have one copy of the allele—are much less likely to be affected with hereditary hemochromatosis, but still may have some minor symptoms (Ina, André, and Martina 2017). The HFE gene creates a protein called the HFE protein that binds to two different transferrin receptors to form a complex that regulates hepcidin, through a molecular pathway that is not yet entirely known (Traeger et al. 2018). There are many common mutations that can appear in the HFE gene. The C282Y mutation (or rs1800562 SNP) on the HFE gene is the most strongly associated with hereditary hemochromatosis, but others such as the H63D mutation can also sometimes cause it. When the C282Y mutation is present, hepcidin is lower, meaning that more iron is absorbed, more than the body needs. Iron absorption occurs slowly, with even those that are homozygous (having two copies of the mutation) not seeing appreciable symptoms until later in life. Men are more likely to eventually have iron overload compared to women, largely because so much iron is used during women's reproductive years (which we will discuss in Chapters 3 and 4). After menopause, however, women who have two copies of one of these alleles may also begin to show symptoms of iron overload.

The C282Y allele is surprisingly common in humans, with a higher frequency in certain regions of the world. It exists at a high frequency in European-derived populations without being selected out of the population. It is estimated to have an allele frequency of up to 11% in parts of Northern Europe (Heath et al. 2016). While people in some parts of the globe may have mutations in non-HFE genes that are associated with iron overload (Wallace and Subramaniam 2016), there is a cline of high frequency of the C282Y mutation in Northwestern Europe (and descendent populations), decreasing as you move south and east across Europe. Given the harms of iron overload, on the surface it seems obvious that natural selection should be selecting this mutation out of the population. When an allele is causing harm but persists in the population at a high rate (even 0.5% is a high allele frequency from a population genetics perspective), researchers assume an evolutionary force is at work. This logic is not just applied to hereditary hemochromatosis, but also to other single-mutation diseases like sickle cell disease, Tay-Sachs disease, and cystic fibrosis. These diseases are all autosomal recessive, and they also persist in certain populations at a higher frequency than expected given the fitness-robbing effects of the autosomal recessive condition. Individuals born with Tay-Sachs disease, for example, usually do not live beyond the age of two, and those with sickle cell disease and cystic fibrosis have only recently been living into adulthood, and only with significant medical intervention.

When we try to explain the maintenance of these deadly single allele genetic illnesses in populations, biological anthropologists traditionally discuss the evolution of the sickle cell allele. It is the gold standard example of an evolutionary mechanism called "selection for the heterozygote." The sickle cell allele causes red blood cells to be shaped like crescents rather than discs, impeding the flow of blood and oxygen to tissues. Sickle cell crises are extremely painful and potentially life threatening, and those who are homozygous recessive face increased mortality. Those who are heterozygous, carrying one copy of the sickle cell allele, also have some sickled cells and some symptoms but not nearly to the same degree. The cline of the sickle cell allele strongly overlaps with thc distribution of malaria across much of Africa, Mediterranean regions, and parts of India and the Middle East. Geneticists have convincingly demonstrated that human modification of the environment for agriculture supported increases in the population of *Anopheles* mosquitos, the carriers of the *Falciparum* parasite that causes malaria. The *Falciparum* parasite cannot infect sickled cells as well as they can infect normal red blood cells. This means there is mortality pressure on both homozygotes. Those without sickle cell alleles face mortality from malaria (which has a high death rate of about 20% if it is untreated), and those with two sickle cell alleles face mortality from sickle cell disease. This means that heterozygotes, those with one copy of the allele, are selected for even though they can experience some symptoms of sickle cell disease. There is a complex, well-demonstrated interplay between human-modified environments, balancing selection, and the clinal distribution of sickle cell disease; and it seems that hereditary hemochromatosis should have a similar explanation. Several hypotheses have been advanced to explain the distribution and persistence of C282Y, but none are quite as convincing as the evolution of the sickle cell allele. Still, researchers have proposed similar evolutionary hypotheses to explain the persistence of the HFE mutation, given the clear adverse health effects to those that are homozygous for the hereditary hemochromatosis.

One hypothesis for the persistence of hereditary hemochromatosis is a pregnancy protection hypothesis. Human pregnancies use a lot of iron (which we will discuss in depth in Chapter 3), and this hypothesis suggests that the higher iron stores found in heterozygotes help women carry pregnancies to term. The balancing selection in this case balances the higher mortality associated with iron overload later in life with the higher mortality associated with iron deficiency during pregnancy and birth (Datz et al. 1998). The main evidence for this hypothesis is that there are higher iron measures in heterozygous women compared to those without the C282Y allele. However, there is no direct measurement of their survival and fitness given their genotype, which weakens the claim for a selective mechanism. This hypothesis also does not account for the clinal distribution of the C282Y allele—if higher iron is so advantageous for reproductive-aged women, why hasn't the allele spread more widely? The allele likely arose 200–250 generations ago (Heath et al. 2016), meaning it may have had time to become more widespread if it was universally advantageous for pregnant women. If this hypothesis is true, the question should be flipped—why aren't hemochromatosis alleles more common in the population? Further, there is evidence that higher iron is *not* necessarily beneficial to pregnant women, a phenomenon I'll explore more in Chapter 3. The fact that the C282Y allele grew to high frequencies within Northern populations and does not appear to have spread more widely has not been addressed by the typical "prevention of anemia" evolutionary hypotheses, but its location might be a factor in the persistence of this allele.

The strong north-south clinal distribution of the C282Y allele suggests the possibility that climate is the driver of its geographic spread. Heath et al. (2016) suggest that iron is needed as part of thermoregulation in cold climates. They point out that the C282Y allele likely originated around 6,000 years before present, coinciding with the expansion of Neolithic farmers into central Europe. These farmers brought with them a notable change in subsistence patterns compared to the hunter-gatherers that had dominated before. While populations beforehand subsisted on wild foods, including iron-rich meats, the farmers were associated with a shift to dairy and grains, two low-iron foods (and in the case of dairy, calcium actively inhibits the absorption of iron). However, this dietary shift alone was not the entire explanation for the cline of the C282Y allele; the authors suggest that this iron deficiency was especially problematic in damp, chilly climates. Iron deficient individuals are less able to regulate body temperature because iron deficiency interferes with thyroid gland function. The combination of low dietary iron and climate would allow the C282Y allele to be maintained in a climate-related cline across Europe. While this is an interesting hypothesis, it is difficult to test, particularly since (as the authors note) technologies such as iron cookware (2,400 years ago) and modern nutritional science (about 100 years ago) have altered the nutrientscape of Europeans. This also means that the disruptive selective force (iron deficiency) may be less likely to maintain the allele in the current population.

A more recent review focuses on the selective force of diet in maintaining the C282Y allele, particularly in Ireland. The C282Y allele is most frequent in Ireland, with allele frequencies at 10–11%, compared to 1.9% estimated in the global population (Burke and Duffy 2022). The authors of this paper point to more recent dietary stressors in Ireland as potential forces that have increased the frequency of this mutation, particularly the Great Famine that was associated with potato blight in the 19th century. Structural forces assured that the available famine-relief foods, such as corn, were low in iron. Analysis of bioarcheological populations in Ireland during that time demonstrated evidence of widespread nutritional deficiencies and death due to infectious disease. Post-famine, bread and tea became a large part of the Irish diet, which are also iron-poor foods. The authors specifically reference gluten in bread as pro-inflammatory within the gastrointestinal tract, potentially blocking the absorption of iron; in addition to the compounds in bread and tea that block iron absorption. They refer to sources that note that tea in Ireland was heavily steeped and drunk excessively, potentially exacerbating iron deficiency. Through this review, the authors created a rich biocultural explanation for high C282Y gene frequencies in Ireland, noting how society and biology interact to create a particular genetic, and ultimately, physiological outcome. This is a pathway of embodiment of iron status—it demonstrates how structural forces—specifically, poverty and famine—can be influential enough to affect the genetic makeup of a population via evolution (Chapter 5).

The last explanation rests on the relative age of those affected by hereditary hemochromatosis. When an allele exerts an effect on mortality, but the effect occurs near or after the end of the reproductive lifespan, it cannot exert a selective effect. Why? Because the organism has already reproduced, passing the mutation to their offspring before the risk of mortality creeps up with age. Only alleles that have an effect earlier in life can evolve via natural selection. The negative allele must impact mortality while the affected individual is still reproducing. The earlier the allele's effect, the stronger natural selection will operate on it. This is known as the strength of selection—that natural selection is strongest when the phenotype occurs before reproduction but becomes less and less effective against phenotypes the more the organism reproduces. Hereditary hemochromatosis could even be a case of antagonistic pleiotropy, in which an allele with detrimental effects later in life will be selected for if it confers an advantage earlier in life.

Why focus on age? Iron overload becomes a problem, including increased risk of mortality, after the age of 40 or later in both men and women. In fact, some individuals with two C282Y alleles never show evidence of iron overload, meaning that the survival of individuals with hereditary hemochromatosis is likely quite high through the reproductive years. The peak of iron overload occurs near the end of the human reproductive lifespan (especially for women), so it cannot be totally selected out of the population. Low strength of selection and antagonistic pleiotropy do not explain the clinal pattern of C282Y, but they could be alternate explanations to balancing selection. Like all phenomena that have evolutionary hypotheses, more than one (or none!) may have caused the allele in question to evolve to higher frequency. To tease apart these hypotheses, measurements of fitness and mortality relative to each genotype would be the first line of evidence, particularly in conjunction with the specific environmental force being proposed (e.g., climate, pregnancy). For example, a researcher might study the completed fertility of women with and without the C282Y allele using genealogical information and medical records in affected areas like Ireland.

The "cure" for hemochromatosis involves regular blood withdrawals, which help prevent toxicity from iron build-up and alleviates the infection risk. More recently, drugs that replicate hepcidin have successfully been used to manage hemochromatosis-related bacterial infections. It is important to note that these innovations are due to learning the "how" of hereditary hemochromatosis—an excellent reminder of the importance of both proximate and ultimate approaches.

Case 3: Inflammation

The third case of iron physiology from an evolutionary medicine perspective is an interesting one. I consider it a case of host-pathogen co-evolution. As mentioned in the section "Iron Hurts," iron can increase the harms of infection because bacteria need iron, too. Bacteria that instigate host inflammatory responses are especially responsive to iron. Excess host iron is therefore a risk factor for increased infection virulence, increased host inflammation, and increased mortality. However, inflammation is not all bad. It helps clear infection and can help promote healing responses after infectious damage or trauma. It is also a signal to other physiological and behavioral systems that can help respond to infection, including iron-related physiology. In fact, two notable inflammation-mediated mechanisms have evolved to keep iron away from pathogens: sickness behaviors and iron withholding.

Think back to when you've been infected with something, maybe a cold, the flu, or COVID-19. You don't feel well, right? Besides the fever you might have, you may not feel like eating or drinking, staying awake, seeing other people, or moving around very much. You probably wanted to stay in bed, alternating between sleep and watching a screen. While it might be tempting to chalk this up as a response to the damage caused by an infection, your response is actually an evolved suite of behaviors called sickness behaviors. Sickness behaviors are an important part of fighting infection: they allow the organism to rest, heal, avoid infecting others, and stop the onslaught of food-borne pathogens (if you have a gastrointestinal infection, that is). Sickness behaviors are triggered by inflammation and are mediated by the immune, endocrine, and nervous systems, which signal to the brain that it is time to deploy them.

Loss of appetite is a particularly interesting sickness behavior for understanding iron. Why? Because loss of appetite helps prevent the host from eating foods that contain iron. So, the first line of defense against iron-hungry pathogens is to not feed them iron, which you achieve by not feeding yourself. In the case of acute infection, this is a temporary adjustment, meant to work only in the short amount of time you are infected. If behavioral changes are not enough to help stave off the more serious infections, iron physiology has its own evolved mechanism to keep iron away from pathogens.

The second mechanism that helps keep iron from pathogens is known as physiological iron withholding. Discovered by Eugene Weinberg in the 1960s and 1970s, iron withholding is a collection of mechanisms that keep iron from being accessed by infectious pathogens and their siderophores (Weinberg 1984). Iron binding proteins such as transferrin and lactoferrin and iron storage proteins such as ferritin already keep iron away from infectious pathogens. In the face of infection, however, excess ferritin is produced to sequester iron within cells, moving it away from the making of red blood cells (called erythropoiesis), and squirreling it away in the liver. Macrophages that have digested old red blood cells will also sequester iron. Sequestering is mediated by hepcidin; high hepcidin not only prevents iron absorption, it also induces iron sequestration. Iron withholding is part of a bigger form of immunity known as nutritional immunity, in which a set of physiological processes keeps trace minerals away from pathogens. Keeping the iron away from pathogens is key; some have even suggested that slight iron deficiency anemia is adaptive against infectious disease (Wander, Shell-Duncan, and McDade 2009). In the case of iron, much of the iron withholding happens in the gut, where the elevated levels of hepcidin prevent iron absorption. However, this leaves more iron in the gastrointestinal tract for the bacteria there, a phenomenon I will explore more in Chapter 5.

Iron sequestering is a vital functional adaptation, evolving not as a single point mutation but as a complex dance that all human physiologies are capable of. It can also be implicated in an evolutionary mismatch: illnesses with an inflammatory component, such as autoimmune diseases, are prone to long-term presentation of both sickness behaviors and alterations in iron physiology. This can present as a special form of anemia known as "anemia of chronic illness." In anemia of chronic illness, hemoglobin is low and ferritin is high because iron is sequestered. This makes the sufferer appear iron replete, when in fact their red blood cell count is too low. This is why ferritin alone is not a good marker of iron deficiency—inflammation can modify it, so it is not a reliable measure of bodily iron. A more complete suite of biomarkers is needed to unpack the real nature of iron status: hemoglobin, ferritin, transferrin receptors, and hepcidin are all part of the iron homeostatic process. See Table 2.1 for a description of these biomarkers and make sure you refer to it as a guide throughout this book.

Concluding thoughts

Iron physiology and evolution are completely and inextricably linked, a tie that goes back to the earliest emergence of Life. That iron's malfunction can be implicated in human illness should come as no surprise; but the evolutionary connections between iron and illness are less commonly known. Human iron physiology can be best understood through the framework of evolution, helping us understand why iron does what it does—why it is a vital but threatening part of the body. This is the initial groundwork for our understanding of

Table 2.1. Commonly used biomarkers of iron status, abbreviations, and their functions. Adapted from Miller (2016).

Biomarker	Abbr.	Function
Functional iron		
Hemoglobin	Hb	Iron-containing protein that binds oxygen for transport to body tissues.
Hematocrit	Ht or HTC	Volume percentage of red blood cells in blood. Can be an indicator of anemia due to blood loss.
Zinc protoporphyrin	ZPP	Compound found when heme production in red blood cells is inhibited by iron deficiency. Confounded by lead exposure.
Iron Storage		
Serum ferritin	sFn	Protein that stores, transports, and releases iron in a controlled fashion; circulates in serum and is found in the spleen, bone marrow, and liver. Serum ferritin levels are believed to reflect global iron stores in bodily tissues. May be elevated during inflammation.
Soluble transferrin receptor	sTfR	Transferrin receptors are present on cells to aid uptake of transferrin-bound iron. Soluble transferrin receptor is a portion of the transferrin receptor that has cleaved off and circulates in serum. Used to determine iron status for those in inflammatory states when serum ferritin may not be appropriate.
Iron Binding Capacity		
Transferrin	Tfn	Protein that binds tightly to iron to prevent free iron from circulating in biological fluids. Transferrin that is not bound to iron is called apotransferrin.
Serum iron	sFe	Circulating iron that is bound to transferrin
Total iron-binding capacity	TIBC	The maximum amount of iron that transferrin in the blood can carry.
Transferrin saturation	TSat	The percent of transferrin iron-binding sites that are occupied by iron.
Unsaturated iron-binding capacity	UIBC	The maximum unbound sites on transferrin that could be occupied by iron.
Lactoferrin	LF	Iron-binding protein found in abundance in secretory fluids, such as saliva, milk, tears, and respiratory secretions.
Iron Homeostasis		
Hepcidin	—	Peptide hormone that regulates intestinal iron absorption, iron mobilization from storage, and iron recycling.

women's iron status, but many questions are still unanswered, more tesserae should be put in place. Notably, these evolutionary hypotheses explain why humans are vulnerable to disruptions in iron homeostasis, but they do not answer why women are uniquely vulnerable to iron deficiency. To understand women's vulnerabilities, we will need to look deeper: specifically, by interrogating the nature of human reproduction. Human viviparity—internal gestation with a placenta—is a resource intense time for a mother that leans heavily on her bodily stores of iron. Evolutionary theory can help explain why reproducing women are more vulnerable than men to low iron. In Chapter 3, we will examine more of the evolutionary tesserae within this mosaic. As we step toward the scene, we are surprised to see a familiar figure depicted there: me.

3
Her Flesh and Blood
Iron and Women's Reproduction

At one point, we look to a corner of the mosaic, and there I am, depicted as a jumble of tesserae. I'm outside a small classroom with a group of women and their infants. The women's bead-and-metal jewelry adds color and sparkle to the mosaic. The picture is framed by blue sky and red earth. We lean in to see what I am doing.

My introduction to iron metabolism came through a side project during my dissertation research. My dissertation focused on milk immunobiology and its effects on infant growth from a life history perspective, looking for buffering effects of human milk on infant growth among settled pastoralists on Marsabit Mountain in remote northern Kenya. My trip to the remote, dry volcano shield had been short but successful, and I left with a liquid nitrogen-filled metal dewar of milk and saliva samples and stacks of paper, smudged with red dirt and filled with precious data. Women had come, dressed in colorful kangas and beaded metal jewelry and carrying their infants, to a small schoolroom to participate in my research. Along with my main research questions, I had also packed a HemoCue. The handheld plastic red box accepted a small microcuvette, filled with a drop of blood, and spit out a number in response. Women offered their fingers, one by one, for this minimally invasive test, rubbing their hands together to cut the chill of the air so that their capillary blood could flow. You could hear the click of the disposable lancet, the woosh of a tissue sweeping the first drop of blood away, and the slide of the HemoCue's tray. I dutifully recorded numbers on a packet of paper and bandaged fingers.

The women of the Ariaal people of Kenya tended to be lean—a product of heavy physical work and spare nutrition. I knew that lean women were prone to maternal depletion in their postpartum months. Maternal depletion is a pattern of energy depletion seen after pregnancy. Pregnancy, with its high energy demands, extracts a heavy burden from women, especially women like the Ariaal. Depending on their circumstances, women can recover from the depletion during the postpartum period before they become pregnant again. Women who cannot recover completely go into their next pregnancy with less

Thicker Than Water. Elizabeth M. Miller, Oxford University Press. © Oxford University Press 2023.
DOI: 10.1093/oso/9780197665718.003.0003

available energy. When this happens over a women's reproductive lifespan, serious deficiencies result, putting women in danger.

Most work on maternal depletion had been done with energy metabolism, focusing on body fat and biomarkers such as insulin and c-peptide. What the HemoCue showed, however, is that almost a third of the Ariaal women were also anemic. It measured hemoglobin levels in blood. Below a certain hemoglobin threshold, a person could be considered anemic, with insufficient capacity to transport oxygen to bodily tissues. Wanting to expand the concept of maternal depletion to other aspects of nutrition, I figured that hemoglobin, as a very crude measure of iron status, might serve to demonstrate maternal depletion of iron. So, around the edges of my dissertation writing, I set out to test this hypothesis.

I was right. There was evidence that Ariaal women's hemoglobin recovered as they moved further from their pregnancy. Hemoglobin levels improved in women with older infants, peaking at about 15–18 months postpartum. However, there was also evidence of depletion, with women's hemoglobin levels decreasing with each successive birth. A loss about 0.2 g/dL per infant equaled over 1 g/dL by the time a woman had the Ariaal average of 6 children. A woman with the population's mean hemoglobin of 13 g/dL would fall into the anemic range by the time she reached the end of her reproductive lifespan. This effect persisted even after controlling for women's age. Excited to find confirmation of my hypothesis, I wrote up the article and submitted it to the *American Journal of Human Biology* (*AJHB*). After some revision, it was accepted (Miller 2010). I was thrilled to be putting one of my first publications into *AJHB*, the journal read by every human biologist in the field.

There was one catch to this finding, pointed out to me by one of the paper's eagle-eyed reviewers. Ariaal women who had higher hemoglobin were more likely to have had their periods return after giving birth. On the face of it, this seemed wrong. Common biomedical wisdom suggested that women who menstruated had lower markers of iron status, not higher ones, due to blood loss. The reviewer, rather than dismissing a finding that ran counter to received wisdom, instead encouraged me to pursue an evolutionary explanation. The reviewer's prompting launched me into the field of human reproductive ecology.

Iron and human reproductive ecology

Human reproductive ecology is an evolutionary field with a basis in life history theory. It is concerned with variation in human reproductive indicators in an ecological context and explains reproductive variation as adaptation to the environment. Human reproduction has been shown to vary in response to nutrition, stress, physical activity and work, disease, and social circumstances. It is studied using population-based approaches to reproductive life events, such as timing of puberty and menopause, completed fertility, duration of pregnancy, lactation and interbirth intervals, mate choice, mating and parenting effort, and reproductive hormone levels. Human reproductive ecology uses a life history framework, in which an organism's time and energy is physiologically allocated to biological events across the lifespan to maximize fitness in their environment. Therefore, human reproductive ecology has a strong bent toward understanding how energy metabolism affects the timing and outcomes of reproductive events. Maternal depletion fits neatly into this framework.

Maternal depletion is primarily an energetic hypothesis. It stems from the delicate balance women's physiology faces between expending energy to support current offspring versus the energy that might be needed for future offspring. During the postpartum period, women's bodies attempt to recover the energetic stores they have expended during pregnancy. Women with more energetic reserves, in the form of fat stores, may take less time to recover from giving birth. Women in worse shape may take longer, increasing the time between births—what we would call a longer interbirth interval. These differences are ecological—dependent on the complex niche in which populations are embedded. In contexts with adequate nutrition and low infection, for example, women's bodies have plenty of energy available to use in their metabolisms; and in others, energy may be in a deficit. Energy balance, then, can only be understood within the human environment, with women's metabolic and reproductive physiologies facing different energetic stresses and buffers depending on the context of their lives.

The ovaries are sensitive to energetic fluctuations. When a woman's body has recovered enough energy to reproduce after birth, she begins to ovulate. However, sometimes this system is not perfectly attuned—women's ovaries may ovulate before their bodies have fully recovered their energy. This leads to maternal depletion. In life history theory-speak, we would call this a trade-off between current and future reproduction. Women who reproduce before their body is fully recovered trade-off their future reproductive ability in favor of their current reproduction. Maternal depletion of energy can result from this trade-off, sapping women's bodily reserves with each pregnancy (Shell-Duncan and Yung 2004).

Life history theory is an evolutionary framework that is concerned with time and energy, trading off in a delicate homeostatic balance across the lifespan. Most biological anthropologists using life history theory have therefore been concerned with energy homeostasis, whether in the form of fat reserves or biomarkers of glucose metabolism. I propose a "next step" for life history theory: to include other forms of metabolism under the umbrella of "energy." Iron metabolism is an excellent candidate for consideration: iron metabolism is older than glucose metabolism and is necessary for a wide variety of metabolic processes throughout the body (including glucose metabolism!). That means we can reconceptualize life history theory as a theory that explains trade-offs in time and *resources*, rather than *energy*, where resources can refer to any bodily substrate involved in metabolism. This would allow life history theory to be used more broadly and can increase the ability of evolutionary theory to explain physiology.

With this in mind, I will discuss human reproductive ecology using iron status as the limiting resource on women's reproductive physiology. I will use a cross-population approach to capture how iron status trades off with reproduction in a wide variety of ecological contexts. There is a wealth of cross-disciplinary research on iron status in women; however, biological anthropology is uniquely positioned to provide this tessera: an evolutionary perspective and ecological contextualization to explain the variation in iron status within and between populations.

Fecundity

The interesting menstruation results I found can be explained in terms of human reproductive ecology. The sharp reviewer pointed me toward a paper I had not seen—a paper with a bold title—"Menstruation Does Not Cause Anemia"—by Kate Clancy and colleagues (Clancy, Nenko, and Jasienska 2006). The population in this study was women from rural Poland—a much different ecology than the women in my study in rural Kenya. The researchers studied endometrial thickness during the luteal phase of the menstrual cycle, a proxy of menstrual blood loss. They found that hemoglobin and red blood cell count, both markers of anemia, were higher in women with thicker endometria. This result ran counter to established biomedical belief that a thicker endometrium, which would shed more lining and more blood during menstruation, should result in higher levels of anemia—and blew up this widely held belief. It showed that a healthy menstrual period meant something entirely different than being drained or left weak from loss of blood. Instead, it meant a woman whose reproductive system was ready to reproduce.

Clancy's results gave my Kenyan results context. Kenyan women began to menstruate only after their bodies recovered enough iron to do so. Their iron metabolism had enough time during the postpartum period, when they are producing milk with low (but highly bioavailable) levels of iron, to recover their iron status. The conclusion here is that iron works similarly to glucose in its ability to show when a woman's body has recovered enough to have their next pregnancy. Once you have enough iron, the body signals that it is ready to reproduce again by having a menstrual period. This readiness to reproduce is a concept known as fecundity.

In contrast to fertility, which is the number of offspring produced by a woman, fecundity is a bit more difficult to define in human reproductive ecology. How do you measure a woman's fecundity? Human biologists rely on a variety of direct and indirect measures to approximate fecundity. First, a major element of fecundity is ovulation. While measuring the release of an egg can be difficult, human biologists have measured hormones that are associated with the release of an egg, such as luteinizing hormone or follicle stimulating hormone. They have also measured hormones associated with her number of remaining eggs, such as anti-Müllerian hormone. However, fecundity is not just associated with ovulation. An egg must become fertilized, it must implant into the endometrium, and the pregnancy must be supported by the woman's body. Endometrial health and thickness and supporting hormones, such as progesterone, play a significant role in supporting implantation and early pregnancy. And a woman's condition, such as her nutritional status or level of stress, plays a role in her body's ability to maintain her pregnancy. These, and hundreds of other physiological factors, contribute to fecundity. The ability to have a menstrual period is an outward sign that a woman is fecund.

The evidence was starting to point toward iron as a crucial resource in the maintenance of women's fecundity. The remaining literature, however, was scarce on a possible relationship between iron deficiency (with or without anemia) and the ability to reproduce. In a letter to the medical journal *The Lancet*, a group of physicians found that seven women with fertility issues conceived fairly rapidly when treated with iron to restore their ferritin levels. While this was not a systematic test of a hypothesis, the authors suggested that there might be a relationship between iron stores and the ability to become pregnant (Rushton et al. 1991). A later study of iron supplementation in women found a relationship between the consumption of iron, either via non-heme iron or supplements, and a lower risk of ovulatory infertility (Chavarro et al. 2006). There was also some suggestion of a proximate mechanism: ovaries contain and express transferrin receptors, indicating that ovarian function is somehow sensitive to iron (Briggs et al. 1999). At the time of my Ariaal paper in 2010, these few studies supported the idea that low iron might be associated with fecundity.

Since then, however, a growing number of studies have begun to interrogate the relationship between iron deficiency and fecundity. An experimental study among rats found that female rats manipulated into iron deficiency via diet had a lower conception rate, disrupted estrus, and delayed conception compared to rats with a normal diet (Li et al. 2014). A study of women who had been referred for recurrent pregnancy loss found they had lower ferritin levels compared to women with no known fertility problems, and an inverse relationship between ferritin levels and number of pregnancy losses (Georgsen et al. 2021). However, they found that ferritin was not associated with ability to conceive or time between study enrollment and conception. Still, the authors concluded that low iron stores were associated with a greater degree of reproductive disturbance and that more exploration of the topic was warranted.

These studies were promising, and I wanted to learn more. Since one of the goals of human reproductive ecology is to assess how evolutionary principles are expressed in a variety of contexts and populations, I decided to look at birth interval and hemoglobin among the Tsimane' people of Bolivia, as a test of the relationship between fecundity and iron (Miller and Khalil 2019). The Tsimane' Amazonian Panel Study, or TAPS, was a multi-year study designed to study the biological, social, and economic changes among the Tsimane' people over a nine-year period. The Tsimane' people are an indigenous rural group of foragers and horticulturalists, participating to varying degrees in the greater Bolivian market economy. For my purposes, the data set contained hemoglobin measurements for at least two years and birth intervals of women in the study. This was also a population that did not make wide use of contraceptives, meaning that the variation in interbirth intervals would not likely be due to family planning efforts. After reconstructing the birth intervals and hemoglobin levels for 116 women, my then student Maie Khalil and I evaluated the question: how do women's hemoglobin levels post-birth affect the length of time between that birth and the next one? The interbirth interval was our measure of fecundity—how long it took for a woman to become pregnant again after she had given birth. And hemoglobin level was our rough indicator of iron status. We expected, given the results from the Ariaal people, that women with low hemoglobin would have longer interbirth intervals than women who had higher hemoglobin levels. We used a survival analysis, which looks at which factors predict the amount of time it takes to reach an event (in this case, the event was birth). And—what we found was not what we expected to find.

We found that women who had higher hemoglobin levels, interestingly, had longer interbirth intervals than women with lower hemoglobin levels.

You can see the survival curve in Figure 3.1, for high, medium, and low hemoglobin levels. On the x-axis, you can see the number of days between births. On the y-axis, you can see what is called the "hazard rate." This is the daily possibility of the birth "event" happening. You can see that the peak hazard of birth is approximately 1,000 days for women in the middle hemoglobin category, and slightly later for the low hemoglobin category. This means that these women have the highest chance of birth happening around 1,000 days after their previous birth, meaning that they become pregnant again when their child is about 2 years of age. This number puts the population squarely within the WHO's recommendation for healthy pregnancy spacing.

The high category was surprising. Rather than a hazard "peak" like the low and medium groups, there was a low hazard rate with a long tail for this group.

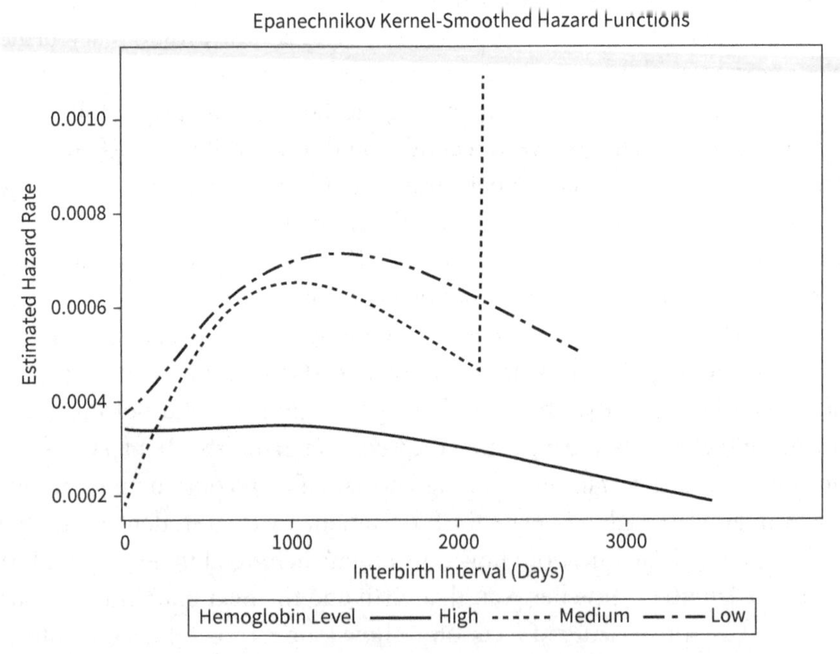

Figure 3.1. Hazard rate (daily risk of birth) of subsequent birth among Tsimane' women by hemoglobin levels: high-, medium-, and low-hemoglobin strata. This figure shows the hazard of giving birth after a previous pregnancy. The vertical line in the medium stratum is an artifact of one participant in that group with a long observation time. For women with medium and low hemoglobin, the highest hazard of giving birth is about 1000 days after the previous birth. For women with high hemoglobin, the hazard of subsequent pregnancy remains low without a "peak," indicating a subsequent pregnancy is less likely for this group.

Source: Miller and Khalil (2019), made available by CC-BY 4.0 (http://creativecommons.org/licenses/by/4.0/).

One way to interpret this information is that women with high hemoglobin have a lower chance of giving birth, and when they do give birth, the birth is likely to happen much later. This result was certainly a shock. We had made no explicit hypothesis about high hemoglobin, but we had at expected that low hemoglobin would have increased birth intervals, pushing the peak birth hazard later for low hemoglobin women. We could explain that lack of finding in part because very few women in the sample were severely anemic—the low hemoglobin category was mostly women with mild anemia. Instead, we see that high iron, a resource necessary for successful pregnancies, was actually associated with lower fecundity. We had to look further to explain these findings.

We found, to our surprise, that research suggests that too much iron might be harmful to the developing embryo. Scientists have long understood that too much iron is harmful to the human body—necessitating the evolution of delicate iron transport systems (Chapter 2). Gene Weinberg, a scientist primarily interested in the regulation of iron, noted that oxidative stress precipitated by iron's reactions was damaging to bodily tissues, and that the embryonic period is the most sensitive period to teratogenesis. Teratogenesis is the process through which congenital defects are produced on embryos and fetuses. Weinberg hypothesized that iron was a teratogen, one that could induce serious harm in the developing embryo. He pulled a few different lines of evidence together to support his hypothesis (Weinberg 2010).

First, he noted that women's iron absorption is lower during the first trimester than pre-pregnancy or in later pregnancy. The amount of iron absorbed during the first trimester was even lower than normal iron losses during this time. The first trimester is the embryonic period, when cells are differentiating into different cell lines, creating the head, torso, limbs, and organs. Second, he pointed out that fluid in the coelum, the cavity in which the embryo rests, has significantly lower iron than is found in maternal serum and higher levels of iron-binding proteins such as ferritin. This suggests that iron is being withheld physiologically from the embryo. Third, the amount of oxygen found early in placental development is much lower than when the placenta is fully formed at 12 weeks' gestation. This means that the amount of oxygen available to create free radicals with iron is significantly lower during the time when the embryo is developing organs. Fourth, iron induces nausea and vomiting—the famous nausea and vomiting during pregnancy, or NVP—whose peak effect is during the first trimester. This is hypothesized as being to get rid of iron that might harm the fetus. While the evolutionary importance of NVP has been debated by anthropologists (Fessler 2002), the debates have largely centered around it protecting mothers and infants from harm. While it may not be the only harmful substance causing NVP, iron is probably one of the culprits.

Finally, Weinberg points to a single mouse experiment demonstrating the teratogenic potential of iron. In this experiment, the experimenter (Kuchta 1982) injected pregnant mice with iron during days 6, 7, 8, or 9, of their pregnancies to be compared with a control group which was not injected with iron. The experimenters found that the offspring of mice injected on day 8 or 9 were more likely to have congenital defects, with most of the defects affecting the brain, spine, and ribs. Days 8 and 9 are what we would call a critical window, or a period in development in which an organism's development is especially sensitive to outside factors. In mice, days 8 and 9 occur during the embryonic period, approximating the latter half of the first month of pregnancy in humans. And while Weinberg does not note this in his review, the mice in all iron treatment groups had significantly higher rates of embryonic resorption, leading to fewer fetuses at term. This means that mice who were treated with iron were more likely to have congenital defects or to experience pregnancy loss (with the embryo being reabsorbed by the mother's body). This underscores the risk to fecundity of too much iron during pregnancy, especially early pregnancy.

In the mouse experiment, it is unclear how much iron was given to the pregnant mice relative to their size and recommended intake. The iron (in the form of ferric gluconate) was also injected and not given as a part of the diet. Injecting iron overrides the evolved systems of iron homeostasis, pushing the iron levels of these pregnant mice far beyond what their bodies would normally ingest. When pregnant mice are given foods with high bioavailable iron, they are shown to have few differences with other iron-replete but non-supplemented mice on metrics such as mother weight, offspring viability, and offspring length, weight, and brain iron (Hubbard et al. 2013). Instead, mice who were experimentally induced to be anemic before pregnancy had lower weight, fewer viable offspring, and lower fetal length, weight, and brain iron. The mice with the high bioavailable iron diet had higher liver iron stores compared to replete, non-supplemented mice. Given that higher iron storage does not affect the pregnancy (positively or negatively), it is possible that important stopgaps exist when excessive iron is consumed via diet during pregnancy (NVP, for one), preventing excessive absorption (but not storage) of iron. Another possibility is that the embryo may be sheltered from iron while other organs, such as the liver, absorb the excess. To my knowledge, this hypothesis has not been systematically tested.

Given the circumstantial evidence, excess iron as a teratogen is an intriguing hypothesis, one deserving of more study in a human model. There are also alternate hypotheses as to why high iron could be associated with reduced fecundity, such as inflammation, hereditary hemochromatosis, or universal poor bodily condition where high iron is part of, but not totally, the reason for altered reproductive outcomes. Untangling these threads would not be easy to do—the suspected period of iron teratogenesis may occur before women are aware that they are pregnant—but it would be an interesting area of research.

The results, while seemingly contradictory, suggest an iron optimum for fecundity. Too high, and excess iron may damage the embryo. Too low, and the mother's condition may be too poor to sustain a pregnancy. Low iron is probably bad for embryos too—besides the mouse experiment with high iron diets, a more recent mouse experiment, one with experimentally induced iron deficiency, has implicated *too little* iron as a factor leading to congenital heart defects in embryos (Kalisch-Smith et al. 2021). This is the beginning of an intriguing area of study, that iron optimums (and variation around them) have real consequences for women's reproductive fitness. Pregnancy offers more evidence that iron optimums dictate women's reproductive lifespans.

Pregnancy

I have mentioned already that pregnancy depletes iron from women. Where does it go? There are a few culprits, but the major cause of iron depletion is the fetus itself. During the first trimester, iron requirements for the embryo are low. However, the transfer of iron from mother to fetus begins to ramp up during the second trimester, and it is estimated that women need to absorb 6 mg of iron per day—almost 5 times more than non-pregnant women absorb (Institute of Medicine (US) Committee on Nutritional Status During Pregnancy and Lactation 1990). About 350 mg of iron are transferred to the fetus and placenta, 450 mg of iron are required to increase red blood cell mass, 250 mg of iron are lost via blood at delivery, in addition to normal physiological losses of 240 mg during the pregnancy period. Overall, about 1300 mg of iron is required for the average pregnancy, with permanent bodily losses of 840 mg. This is well over one-third of the average total body iron content of 2200 mg!

While iron that is allocated to increasing the number of red blood cells can be reabsorbed when blood volume decreases after pregnancy, iron transferred to the fetus or lost during childbirth cannot. And low iron, combined with partum and postpartum bleeding, can put women at risk for severe illness or even death. A woman's iron does not just serve her offspring—it also protects her life. And since iron is a highly regulated, limited resource, women's physiology must balance the transfer of enough iron to grow a healthy child with keeping enough iron to protect the mother.

This is another life history trade-off. In life history theory-speak, the iron in women's bodies is the limiting resource, which must be allocated via metabolism to either (1) keep her alive or (2) assist her body in reproduction. If her physiology does not keep enough iron for herself, a woman risks ill health and potentially death. If her physiology does not send enough iron to her fetus, her offspring is at risk for poor health outcomes. This means that women's physiology must balance these two competing interests: the maintenance of the body (somatic maintenance) and reproduction. Of course, this trade-off will occur in the context of a woman's available body iron and her environment (dietary, infectious, and otherwise). This trade-off is also sensitive to ecological conditions, just as reproductive ecology predicts.

At the current time, there are few to no studies using a life history theory/reproductive ecology framework to predict how iron trade-offs manifest during pregnancy. How would iron trade-off between somatic maintenance and reproductive effort, and how could these trade-offs be measured? In this case, a thought exercise will do—we can create predictions from theory and test them from the literature. If a woman's physiology has limited iron, her pregnancy may be modified in certain ways to manage the low iron. Her body might (1) limit the length of a pregnancy by giving birth early (remember, time is also a limited factor in life history theory) or (2) it might limit the amount of iron delivered to the fetus, leading to poorer fetal outcomes but potentially preserving her limited iron. Is there any support for this in the literature?

A review of iron status variables and birth outcomes across multiple studies found that rates of preterm birth, stillbirth, and small-for-gestational age status is higher among women with iron deficiency anemia, but only if women's iron status is low at the start of pregnancy (Dewey and Oaks 2017). Low iron is not strongly associated with these outcomes if a woman's anemia shows up in the second or third trimester. This is, in part, due to the nature of the biomarker used. Because women gain excess fluid as their pregnancy progresses, it dilutes their blood, making hemoglobin an unreliable measure of iron status during later pregnancy. Even so, in the studies using ferritin as a marker, there is evidence that low iron (as ferritin) is associated with preterm birth and small-for-gestational age when it is low in the first trimester. This review provides mixed support for the idea that women trade-off timing of birth or offspring size based on their iron levels. This is a question deserving of more controlled studies of iron status and absorption, inflammation, and birth timing.

Interestingly, there also appears to be a risk from having too-high iron during pregnancy as well. In the same review, Dewey and Oaks found that there was a "U-shaped curve of birth outcomes," where the risks of preterm birth and small-for-gestational age were higher for those women who had both low and high iron status. Specifically, there is reasonable evidence that high iron in the third trimester (as ferritin) is associated with preterm birth and small-for-gestational age and some evidence that supplementing iron to non-anemic women during the second and third trimesters also resulted in a greater likelihood of preterm birth and small-for-gestational age. It is not easy to disentangle this from a life history perspective—perhaps women who have high iron by the end of their pregnancies have physiologies that are not effective at transferring iron to their infants, ending the pregnancy early and leading to smaller infants. However, this is an untested hypothesis.

Contrast this with the effects of high iron in the first trimester. There is evidence, among a population of Chinese women, that those who had a miscarriage (also called a spontaneous abortion) had higher iron during the first trimester compared to women in their first trimester who did not have a miscarriage (Guo et al. 2019). This work supports Weinberg's hypothesis—that higher iron, at least during the first trimester, is associated with worse birth outcomes than lower iron. However, the results show more. This study also looked at the relationship between ferritin and hepcidin in three groups of women who did not have a miscarriage—a group of women who were not pregnant, another in their first trimester, and a third in their second trimester. As expected, they found a linear positive relationship between ferritin and hepcidin—as ferritin increases, so does hepcidin, a nod to the absorption-blocking function of higher hepcidin levels. Figure 3.2 shows the slope and intercept between ferritin and hepcidin in each of the three groups. Of note, while the slopes of the relationships are the same in each group, the intercept—the place where the line crosses the y-axis—is much higher in the first trimester compared to non-pregnant or second trimester women. Hepcidin maintained the positive linear relationship with ferritin in each trimester, meaning that it was responding to physiological iron levels, but the "set point" for the hepcidin appears to be higher in the first trimester than at any other point. Even more intriguing, inflammation, a typical cause of high hepcidin, was not the driver of this upward shift. Clearly, iron absorption is limited in the first trimester via an increase in hepcidin levels. However, it is much less clear what signals hepcidin to do this, or how it affects life history trade-offs.

Given these contrasting results, it could be that women with dysregulated iron homeostasis face adverse birth outcomes, with the type of outcome dependent on the dysregulation (high or low) and the trimester in which the

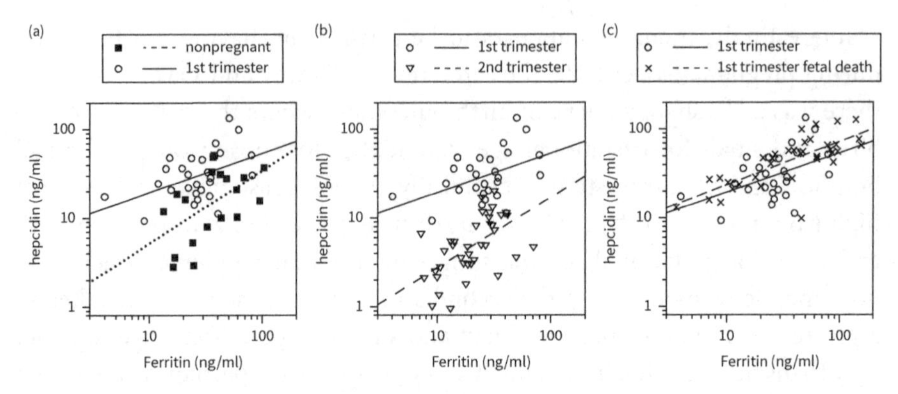

Figure 3.2. The relationship between hepcidin and ferritin in physiological and pathological pregnancy. Scatter plots (log-log scale) and regression lines of hepcidin vs. ferritin in: (A) nonpregnant women (solid squares and dotted line) compared with the healthy first trimester (open circles and solid line); (B) healthy first trimester (open circles and solid line) compared with healthy second trimester (triangles and dashed line); and (C) healthy first trimester (open circles and solid line) compared with first trimester fetal demise (crosses and dashed line).
Source: Reprinted from Guo et al. (2019) with permission from John Wiley and Sons.

dysregulation occurs. Therefore, the trade-offs faced between iron status and reproductive outcome will vary by trimester, or at least may be different between the first trimester and the latter two. On the public health side, there may be profound implications for currently accepted wisdom on iron supplementation during pregnancy. Instead of high, unrelenting iron supplementation for at-risk women during pregnancy, iron supplementation should be reconsidered in the first trimester of pregnancy. In addition, non-pregnant women should be assessed and supplemented if necessary, so that they do not become pregnant while suffering from iron deficiency. Unfortunately, current iron campaigns tend to focus on women who are pregnant, and by then it may be too late to stop the effects of low pre-pregnancy iron.

Pregnancy's depletion of iron is subject to life history trade-offs, then, but the nature of these trade-offs very much depends on their timing. Before birth, too-low iron is considered a risk to the pregnancy—not enough iron means that women face more severe trade-offs between somatic investment and reproduction. During the second and third trimester, the trade-offs are similar, manifesting as risk of preterm birth and small-for-gestational age for the offspring and higher risk of poor birth outcomes for the mother. The first trimester, then, is the mystery—while the risk of high iron may be explained by Weinberg's protective hypothesis, it is unclear why hepcidin shifts the way it does during that time, or what signal triggers the shift. These triggers are likely related to a physiological process during the first trimester, but their nature is unknown. While there are substantial gaps in our understanding, reproductive ecology can still guide our understandings of the trade-offs, risks of dysregulation, and outcomes of iron homeostasis in pregnant women.

Fertility and the postpartum

Now that we know how pregnancy depletes iron from women, let's return to the concept of maternal depletion. I have already related how the Ariaal people showed evidence of permanent maternal depletion of iron, with their hemoglobin levels dipping lower with each pregnancy. This depletion is the long term-consequence of a life history trade-off between somatic maintenance and reproduction. This trade-off is not necessarily bad if you have enough somatic resources to either maintain or replete iron lost from pregnancy. Is that the case for the Ariaal?

Evolutionary trade-offs operate given the bodily resources that are available. If a body has fewer resources during pregnancy (in this case, iron), the physiological trade-offs will be more severe. The results might be preterm birth, a smaller or anemic infant, or severe iron deficiency that threatens maternal well-being. In the case of Ariaal women, it is difficult to tell with the available information whether they have enough iron. Since my original dissertation research was not designed to answer these questions, there are inadequate measures of maternal dietary intake and body iron. My data, and previous studies by other anthropologists, did seem to indicate that typical foods for Ariaal women were not iron-rich. Meat and blood, two heme-rich sources of iron, are either too expensive for the average family to consume regularly or are designated as food for men (Chapter 5). Women are more likely to drink milk, which blocks absorption of iron. And the staple food, ugali, is a maize porridge with little available iron. The lack of data makes it difficult to assess the degree of trade-off happening to Ariaal women, although it is reasonable to assume that dietary iron may not be sufficient in many cases.

Ariaal women's depletion could be the result of the very real dietary deficiencies they face. On the other hand, iron depletion may reflect more subtle realities of the trade-off between somatic maintenance and reproduction: the trade-off between current reproduction and future reproduction. In life history theory, organisms' physiologies try to find the most adaptive way to balance their resources over their lifespan. This means that trade-offs don't just happen over one pregnancy but instead are portioned out over the reproductive years (and beyond). Because an organism will die at some point in the future, possibly the near future, the current reproductive moment is always favored. However, because organisms are hypothesized to reproduce as adaptively as possible given their circumstances, their physiology is generally disposed to make sure that future reproduction is also feasible. In situations where risk of mortality is higher—where death may happen sooner—trade-offs make current reproduction highly favored over the possibility of future reproduction. Shifting between current and future reproduction is mediated by a variety of developmental and physiological mechanisms, including hormones like cortisol, immunological processes like inflammation, or epigenetic mechanisms that change the expression of the genome. In situations where current reproduction is highly favored, there will be less investment in the body's somatic maintenance to funnel resources to reproduction. In situations where future reproduction is the best strategy to maximize fitness, the body's physiology is geared toward somatic maintenance—making sure the organism is up to the task of reproducing in the future.

Humans, with our long lives and low adult mortality rates, tend to favor future reproduction more than most other mammals. This means that typical adult human physiology is geared toward making sure reproduction can happen over the entire reproductive lifespan. However, current reproduction is always at least slightly favored. There is always the risk that you may die during your reproductive lifetime, meaning physiology shouldn't skimp on effort in the here-and-now. This may vary within human populations depending on the adult mortality rate. For example, there is some evidence that people who grew up in dangerous situations may have accelerated reproductive timing, including earlier age at menarche (Amir, Jordan, and Bribiescas 2016)—meaning that their biology is favoring current reproduction. However, even populations who have a low mortality rate have a small bias toward current reproduction.

Given that there is a spectrum between "slightly favoring current reproduction" and "maternal depletion," it's not clear what Ariaal women's parity-induced hemoglobin decline really means. The relative population mortality in Kenya offers a clue. According to the World Bank, the adult mortality rate[1] for Kenyan women was nearly 295 per 1,000 adult women in 2008, the year I did my fieldwork (World Bank 2021). While I do not have mortality rates for Ariaal women specifically, it is likely that it is even higher, given their reduced access to healthcare and marginalized status within Kenya. By contrast, the mortality rate for adult women in the United States was about 79 per 1,000 women. There is a clear difference in adult mortality between the populations,

[1] The World Bank defines adult mortality rate as the probability of dying between the ages of 15 and 60.

indicating that Kenyan woman may be facing a greater tendency to favor current reproduction compared to women with lower mortality rates, such as in the United States. I wanted to look at the degree of this difference, so I decided to test the parity-iron relationship among U.S. women.

I examined maternal loss of iron with increasing parity in U.S. women. Most women in the United States, at least on paper, ought to have the resources to maintain sufficient iron stores for their future reproduction. The National Health and Nutrition Examination Survey, or NHANES, is a gold mine of health and nutrition information about the people of the United States. This sample, publicly available to those in the U.S., generates results that accurately represents the demographics of the country. In theory, the United States is a well-nourished country, with a small percentage of people having nutritional deficiencies. However, the data showed that 6.9% of non-pregnant and 29.1% of pregnant women were anemic in the NHANES, compared to 28.0% among postpartum Ariaal women. In addition to hemoglobin, the NHANES had even more iron status information, such as ferritin and soluble transferrin receptor (sTfR). Using these biomarkers, I was able to ascertain iron deficiency as well. In this case 9.8% of non-pregnant and 25.4% of pregnant women were iron deficient (Miller 2014).

Using this data, I found that there was, in fact, a decline in hemoglobin with parity (Miller 2014). This decline was seen with ferritin levels, too, meaning that the hemoglobin decline was due to a decline in iron levels within the body, rather than another cause. The slope of the hemoglobin decline, however, was less severe than seen in Ariaal women: Kenya $\beta = -0.20$; United States $\beta = -0.08$ (Figure 3.3). This lends support to the hypothesis that even in low mortality conditions as in the United States, there is still bias toward current reproduction. And it shows that the degree of bias toward current reproduction is noticeable when you look across populations. There is a caveat to the U.S. results: not all women in the United States experience the exact same mortality rate. In fact, there are serious disparities in iron status due to race and ethnicity, reflecting both a history and a present of extreme racial inequality. I will discuss these disparities in Chapter 6.

There are other caveats to the comparison between Ariaal and U.S. women: U.S. women have lower overall fertility, having fewer children during their reproductive lifespans than Ariaal women. The consequences of this are that U.S. women may not have the opportunity to deplete their iron stores as much as the Ariaal, because they have fewer completed pregnancies. The evolutionary explanations for lower U.S. fertility are best left to another book, but for simplicity's sake, I'd like to point out that fitness, as a means to quantify natural selection, only means that an individual needs to survive and reproduce more

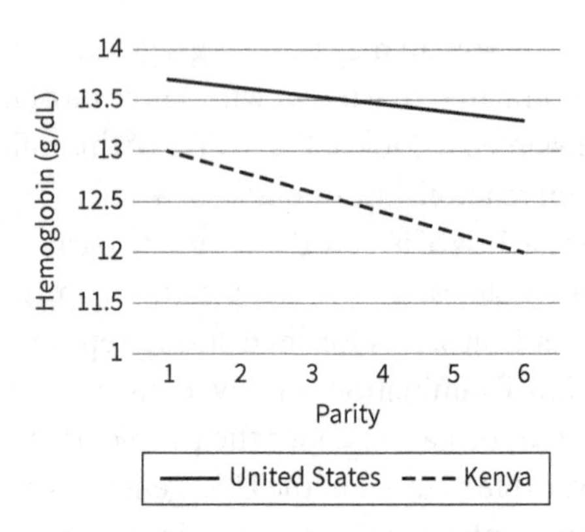

Figure 3.3. Hemoglobin levels by parity in reproductive-aged women, United States (NHANES) and Kenya (Ariaal).

than those around them. Demographically, the United States has undergone a shift toward fewer children; however, there is still variation within the population, fertility-wise. This means that evolution is still at work, albeit with different social, ecological, and environmental constraints acting upon the population.

The evidence from U.S. women shows that a slight decrease of iron with every pregnancy is a feature of life history theory, not a bug. If the depletion does not impact survival and fertility, then from an evolutionary standpoint it is a valid strategy to maximize reproductive output. This does bring up an important distinction though: the differences between health outcomes and evolutionary outcomes. If iron depletion pushes a woman's hemoglobin below a certain cutoff point, her doctor might diagnose her with anemia and treat her with iron supplementation. However, if this anemia does not affect survival or fitness, it won't be selected against—meaning that from an evolutionary perspective there is no effect. Where health outcomes and evolutionary outcomes overlap, then, is when poor health impacts survival and fitness, such as when birth outcomes result in a risk of mortality for the mother or baby—postpartum bleeding, for example, is made much worse by severe iron deficiency (Kavle et al. 2008). Iron deficiency, then, is a place of poor health and of reduced evolutionary fitness, meaning that we can see maternal depletion as a continuum of both evolutionary outcomes and health outcomes. These two perspectives converge on the risk of severe health outcomes. Where they diverge, each can learn from the other—from the evolutionary perspective, a mild anemia may be a normal feature of reproductive-aged women's physiology; from the clinical perspective, anemia is a reminder to dig deeper into the outcomes of altered physiological states, to confirm where this physiology switches from normal to detrimental to survival.

Conflict

> This section has discussed conflict over glucose levels in maternal blood, but similar conflicts would be expected over minerals and other nutrients if these are in short supply. (Haig 1993 p. 513)

I imagine now that sharp readers have been absorbing the idea of life history theory and are thinking, what about the offspring? Shouldn't they have their own fitness, survival, and trade-offs? Yes, they do! Offspring have their own iron needs, which, from an evolutionary standpoint, their physiologies pursue vigorously. When offspring are dependent on their mother's iron stores during pregnancy or lactation, the offspring's needs may come into conflict with their mother's needs. In evolutionary biology, this would be an example of a parent-offspring conflict.

A parent-offspring conflict is expected to occur any time a parent and their offspring have competing evolutionary interests. It may surprise you to hear this, but each has evolved slightly different fitness goals. A parent, in addition to loving all their children very much, has their evolutionary fitness tied up in the survival and fitness of their offspring—all their offspring, equally. Their physiologies want to devote equal resources to their children across their lifespan. So, their bodies will hold some resources back from the current offspring to make sure their biological capacities can support their future offspring. Sound familiar? This is the trade-off faced between current and future offspring.

Offspring themselves, while certainly also loving their parent very much, have not evolved to share this same goal. Their reproductive fitness is tied up in making sure they, and their genes, survive and reproduce into the next generation. Why genes? Well, by switching focus to the genes (or the "unit of inheritance"), scientists can explain—evolutionarily—why altruism might exist. That is, organisms are motivated to improve not just their own fitness and that of their offspring, but also the fitness of everyone who shares copies of their genes. This means that you will help your own fitness first, because you are 100% related to yourself. You'll assist the fitness of your own offspring, being 50% related to them. You'll even help your siblings (25% for half siblings and 50% for full siblings) and their offspring (12.5–25% shared genes). Now, this is a simple and simplified model of evolution. We know that evolution works through multiple levels of biological organization, from gene to cell to organ systems to individuals. However, a gene-focused view can be useful to predict certain phenomena, particularly relating to the conflicting actions of parental and offspring genes.

Parents and offspring conflict on the amount of resources they have evolved to provide to their relatives. Parents' evolved biologies invest in their offspring mostly equally to maximize their fitness—the direct fitness from their offspring being the biggest contribution of their genes to the next generation. From the evolved point of view of the offspring, however, the parent's equal spread of resources does not maximize the offspring's fitness. Since they are only partially related to their siblings, an individual offspring's biology would seek more investment in themselves and less investment in their siblings. When a parent and their offspring's interests interact, there may be conflict. This can manifest as behaviors, physiological adjustments, or even illnesses, depending on when the conflict occurs. Conflicts tend to emerge, according to theory, during points in the lifespan where offspring are highly reliant on their parent for survival. That means that pregnancy, lactation, and early childhood are all major windows for parent-offspring conflict.

An example of parent-offspring conflict is genetic conflicts of pregnancy, which incorporates the proximate mechanisms governing the conflict, allowing us to understand how they manifest physiologically. In such examples, the conflict occurs due to the differential expression of maternal genes (which are invested in supporting each of her pregnancies to maximize fitness) and paternal genes (which are invested in supporting their current offspring, since there is no guarantee that his genes will be involved with the mother's subsequent pregnancies). These conflicts play out as a battle over energy, as the mother's physiology dukes it out with placental physiology (representing the interests of the fetus) in the regulation of blood glucose. David Haig, in his work on genetic conflicts, outlines just such a scenario (Haig 1993). Mothers and fetuses share access to the mother's blood supply via the placenta implanting into the uterus. Placental hormones, like human placental lactogen, can manipulate maternal physiology for its own benefit; for example, by opposing maternal insulin production to raise blood glucose (which benefits the infant). Maternal physiology reacts by increasing the production of insulin. For most pregnancies, these opposing physiologies cancel each other out, and the infant is born at a normal size. However, these physiologies are variable and responsive to nutritional conditions, so there may be variation in birth weight. In fact, maternal blood glucose correlates positively with birth weight. If the fetus's genes (via placental hormones) persevere in the conflict, the result is gestational diabetes for the mother and an oversized infant at birth.

Haig points out that there may be conflict over other nutrients if their physiological availability is limited. During pregnancy, iron very much meets this condition, meaning that maternal-fetal conflict over iron physiology is a distinct possibility. The placenta and fetus, like the mother, are also capable of producing hepcidin, the hormone responsible for maintaining iron homeostasis. It appears—on the surface—that maternal hepcidin levels are the factor that predict fetal iron status, rather than hepcidin produced by the placenta and fetus. In multiple mouse models, researchers have found that maternal hepcidin is inversely associated with fetal iron status (Sangkhae, Fisher, Wong, et al. 2020). Remember that the second and third trimesters together are a low hepcidin state, and that when hepcidin levels are low, iron absorption is high. Absorbing more iron benefits both the mother and the fetus, so it is perhaps unsurprising that these interests mostly correspond. However, even when maternal-fetal interests are similar, they are not necessarily the same. Parent-offspring conflict would predict that the fetus would physiologically be primed to manipulate greater iron transfer to itself than the mother's physiology might typically transfer. However, fetal hepcidin, at least in mouse models, does not contribute at all to maternal iron absorption (Sangkhae, Fisher, Chua, et al. 2020).

The initial evidence suggests that there is not a maternal-fetal conflict over iron. New research, so new that it has only been presented at conferences, appears to have identified a substance tentatively called "hepcidin-suppressing factor." The authors hypothesized this factor after discovering that maternal hepcidin begins to decline before mothers' liver iron stores begin to decrease (Sangkhae, Ganz, and Nemeth 2020). Normally, a decrease in liver iron helps trigger lower hepcidin, indicating it's time for the body to increase its iron stores. The researchers believed that something in the placenta—the "hepcidin-suppressing factor"—may be contributing to the early decrease in hepcidin. They did an in vitro—in a test tube—experiment showing that a not-yet-named protein was expressed from placenta cells, which suppressed the expression of maternal hepcidin mRNA. When hepcidin mRNA is expressed by cells, it means that the hepcidin gene is active—and this factor was successful at blocking it. This means that the placenta might suppress maternal hepcidin to favor absorption of iron for the fetus, even if the mother's iron stores are sufficient. It's possible there is conflict over iron after all—but much more research is needed to determine the nature of the protein and how it might work in vivo—in the body of mothers.

Parent-offspring conflict is at its height during pregnancy but can appear at other points post-birth as well. Lactation is another potential point of iron conflict between mother and infant. How do mothers' bodies manage to recover iron between births? If they are so busy feeding their infants iron-rich milk—the only food young infants evolved to eat—how can they recover their own iron status for their next offspring?

It turns out, there may not be a conflict during nursing—at least over milk iron! Human milk iron is shockingly low compared to the iron needs of young infants. This means women are not mobilizing bodily stores of iron to feed their infants. However, infants do need iron, as the lack of iron has serious consequences for growth and development, especially cognitive development (Radlowski and Johnson 2013). How do they get the iron, if not from their mothers' milk? Well, pregnancy transfers enough iron to protect a full-term infant for up to six months after birth. This does not necessarily apply to infants that are preterm, have fetal anemia, or other causes of poor iron transfer during pregnancy. For term infants, there is a small amount of iron that is transferred via milk, through a highly efficient system. The iron in milk is bound to a special protein called lactoferrin. This protein serves two purposes. First, it protects the iron from being used by pathogenic bacteria, protecting the infant's fragile immune system from infection. Second, it makes the iron in milk highly bioavailable, allowing up to 50% of it to be absorbed by the infant gut. The iron in human-manufactured formula, by contrast, is much less bioavailable, around 10–15%, and therefore has higher iron levels to compensate (with concomitant increase in infection risk).

Lactoferrin makes human milk a highly efficient system for the transfer of iron. While there may be other parent-offspring conflicts over human milk, a conflict over iron is less likely—the major conflict is likely to be during pregnancy when high amounts of iron are transferred. Thus, the interbirth interval is a protected time for women to replete iron—there are few reproductive-related demands on her iron except to recuperate her iron stores. One area of future research would be to study the role of hepcidin in women after birth to see the dynamics of iron repletion. This may inform strategies to help women replete iron more quickly and avoid maternal depletion.

Conclusions

Iron is tightly woven into women's reproduction. A woman's fecundity, her pregnancy outcomes, and her interbirth intervals all track with her available iron. The heavy reliance on iron during pregnancy can put women whose nutrition is already marginal at risk for serious health outcomes. The evolutionary dynamics of iron during pregnancy are not well understood, and more work from an evolutionary perspective is needed. However, the preliminary work suggests that women's physiologies make trade-offs in the face of limited iron. The exact nature of potential mother-fetal iron conflict during pregnancy is also up for debate and requires more research to definitively establish. Overall, however, the bulk of the literature shows a pattern of trade-offs, within certain parameters—women's physiology tends to avoid high iron, making women vulnerable to low iron. This vulnerability is not borne equally by all women, which I'll explore in later chapters.

Beyond this, these findings show that fluctuations in women's iron status can be explained using reproductive ecology and life history theory. Reproductive ecology has been used successfully to explain reproductive variation in energetic status (Ellison 2008; Valeggia and Ellison 2009), and now it shows incredible utility to explain iron status as well. This demonstrates two things. One, that iron is an underappreciated life history somatic resource, vital to the functioning of the human body over the lifespan. And second, that life history theory is a robust framework to explain physiological variation across the life course. Therefore, it should be expanded beyond energy metabolism and potentially applied to other physiological processes. Iron is a useful test case, but there are thousands of other metabolisms in the body that are waiting to be explained with an evolutionary perspective.

While reproductive ecology can explain the ebb and flow of iron status across women's reproductive lifespans, it does not necessarily explain why reproductive women's iron levels are lower to begin with—the divergence between women's iron stores and men's iron stores begins around puberty, years before most women become pregnant for the first time. Women's lower iron has traditionally been attributed to menstruation—but is that true? In Chapter 3, we'll explore how menstruation has become the leading explanation for women's iron woes.

Although it is interesting to see me in the mosaic, there is a sense that this image is incomplete. I am even depicted as holding tesserae, as if I participated in its construction. Shrugging, our eyes catch a bright red drop in another part of the mosaic. What will that corner hold?

4

Out for Blood

Iron and Menstruation

We shift our eyes to the new corner of the mosaic, where it depicts the opening of the 1976 horror movie *Carrie*. A young woman, showering after a gym class, becomes panicked when she discovers she is bleeding. Instead of receiving help, Carrie is cruelly mocked by her classmates for not realizing that she is menstruating. She cowers at the bottom of the shower as they throw menstrual pads at her, screaming at her to "plug it up." Red tesserae spot the floor of the shower below Carrie. This pivotal scene marks her frightening transition into womanhood. An apt setup for a horror movie, right?

Those with female reproductive organs have their lives marked by the regular appearance of blood.[1] Even though menstruation is a reality for hundreds of millions of people—mostly women—biomedical science's ideas about menstruation are grounded in myth, making it seem mysterious, uncontrollable, and yes, scary. Menstruation has been viewed as evidence of the weakness of the female sex. The assumptions made about the depleting effects of menstrual blood loss have entrenched this view. Each month, women lose precious iron via menstrual blood, pushing their iron levels down and making them exhausted after exertion. Incapable of running away from predators, women instead are bound to fainting couches and stuck with the (clearly) less strenuous task of gathering berries. If you haven't guessed from my sarcasm and mixed-up millennia, the idea that women would constantly leak iron, putting their survival and reproduction at risk, is not evolutionarily sound. In this chapter, I am going to brush away the dust on this corner of the mosaic, unpacking the myths of menstruation and proposing three possible evolutionary hypotheses to explain reproductive aged women's uniquely low iron status—and in the process, attempt to rehabilitate the reputation of the menstrual period.

[1] For the most part, these individuals identify as cisgendered women, although transmen and nonbinary individuals' lives can also be bound by menstruation. For the purposes of this chapter, I will focus on those with female reproductive organs and will use the shorthand phrase "women," with the understanding that not all who menstruate are women, and not all women (such as transwomen) menstruate.

Thicker Than Water. Elizabeth M. Miller, Oxford University Press. © Oxford University Press 2023.
DOI: 10.1093/oso/9780197665718.003.0004

A lifetime of iron

Reproductive aged women occupy a unique iron phenotype. In Figure 4.1, I have plotted hemoglobin levels and ferritin levels across the lifespan in U.S. men and women from the National Health and Nutrition Examination Survey, or NHANES (Miller 2016). You can see how hemoglobin and ferritin are the same in boys and girls until about the age of 15 or so. At that age, suddenly male ferritin—and hemoglobin to a lesser extent—dramatically

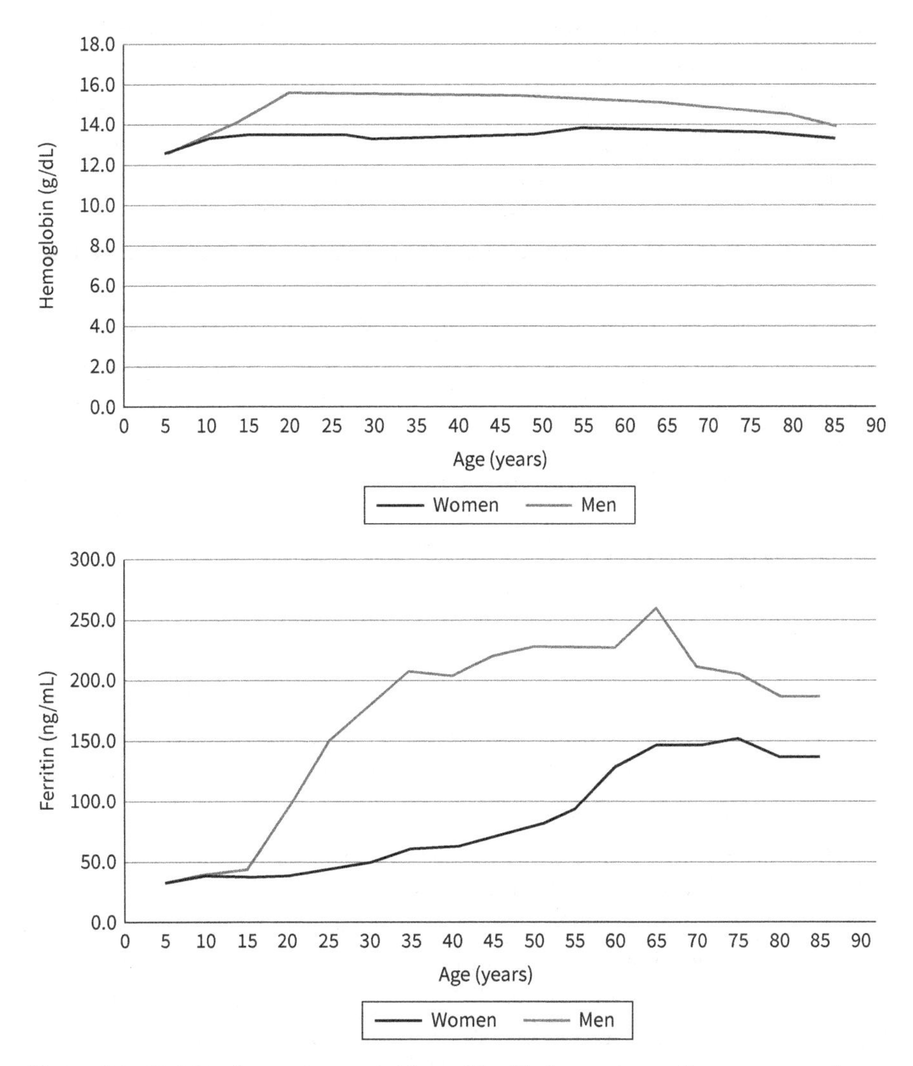

Figure 4.1. Weighted mean hemoglobin and ferritin by sex across 5-year averaged age groups, U.S. NHANES, 1999–2006. The spike in ferritin in men aged 60–65 is mainly due to two men with ferritin levels almost 10 times the mean level.

Source: From Miller (2016) used with permission from Springer Nature.

increases until the age of about 35. From there, male iron stores (ferritin) and hemoglobin levels are higher than women's for their entire lifespans.

Women's patterns are different. Girls' hemoglobin rises slightly through childhood, along with boys, until the mid-teen years. From there, hemoglobin levels remain consistent through the lifespan, holding steady at just over 13.5 g/dL for the average U.S. woman. Ferritin is even more interesting. It does not rise precipitously like men's ferritin, instead increasing slightly though the reproductive years and increasing more dramatically from the age of 55–65, holding steady at a higher level during the post-reproductive years. Researchers explain the patterns of women's iron status over the lifespan by pointing the finger at menstrual blood loss.

Reproductive-aged women menstruate. Menstruation is the beginning point of the menstrual cycle, the monthly cycle of ovulation. During the beginning of the cycle, a single egg-containing follicle in the ovary begins to grow larger and also stimulates the growth of the lining of the uterus, called the endometrium. At the midpoint of the cycle, the egg bursts through the follicle and out of the ovary, making its way sedately through the fallopian tubes to the uterus. If the egg is fertilized, it will divide and implant into the endometrium, beginning pregnancy. If the egg is unfertilized, the endometrium sheds from the uterus (along with a little blood) and the cycle begins again. This cycle is under the hormonal control of luteinizing hormone (LH), follicle stimulating hormone (FSH), estrogen, and progesterone. LH and FSH are pituitary hormones that help stimulate the release of the egg. Estrogen aids in the proliferation of the endometrium during the menstrual cycle. Progesterone, sometimes known as the pregnancy hormone, helps maintain the endometrium until the start of the next menstrual period (or throughout pregnancy). Menarche is a girl's first period. Menopause occurs because ovulation, and thus menstruation, ceases. In women, menarche can happen between ages 10 and 20 years, depending on nutritional and other factors. Menopause occurs at around 50 years of age in women, with modest variation in age at menopause between societies.

The biomedical assumption is that women's iron stores are lower due to the loss of iron with menstrual blood loss. But is this true? Remember, because men have been considered the "default" since the beginning of biomedicine, women's iron stores have been cast as pathologically low. But what if it wasn't a pathology? Would evolution have favored a system that makes an entire sex pathological? To fully consider possible evolutionary hypotheses for women's iron status, we must first unpack the assumptions that cast women's bodies as abnormal.

Menstruation myths and realities

> Thus the woman, in a rather chronic manner, is the theater of these bloody demonstrations. The feelings that the blood evokes are carried within her . . . The woman is therefore, and in an equally chronic manner, taboo for the other members of the clan. (Durkheim 1963 p. 85)

When women present to the doctor with symptoms that might indicate anemia, menstruation is the typical biomedical explanation. You tell your gynecologist you're tired, they test your blood for anemia and tell you to take an iron supplement because period blood loss is what is causing your anemia. You might imagine that this bit of wisdom was arrived at through careful scientific testing. To a degree, this is true. However, menstruation was doomed to be considered pathological by medical doctors from the start.

Anthropologists use the word taboo to describe phenomena that are too repulsive for ordinary people to contemplate. The menstrual period features prominently in the literature on taboos. Apparently, menstruation is repulsive: Menstrual taboos of all stripes have been documented around the world. This taboo manifests in a variety of ways, from separating the menstruating woman from society to suppressing conversation about menstruation. Emile Durkheim, in fact, believed that religion arose in part because of the "repulsing" force of menstruation (Durkheim 1963). Despite the prominent positioning of menstruation as taboo, it is not always all bad. In some societies, separation into women-only spaces offers a relief from the demands of everyday life (Buckley and Gottlieb 1988). Even in the United States, menarche—a girl's first period—is often celebrated with period parties, while at the same time, confusingly, being described as a "curse." Mixed menstrual messaging is a common feature of societies like the United States. These mixed messages have wound their tendrils into the practice and science of biomedicine.

Menstruation is a common biomedical taboo, although biomedical practitioners would insist that it is not. Feminist social scientists have long noted that menstruation was viewed as pathological by the Western medical establishment of the 19th and 20th centuries. Strange (2000 p. 610) notes that in reports on menstruation in the medical journal *Lancet* in the 1850s, "the medical definition of menstruation was almost exclusively expressed in terms of pathology and a failure to reproduce." This belief in the pathological nature of menstruation extended to women's intellect and abilities outside of the home—according to the physician Edward Clarke, education disrupted the normal functioning of the female reproductive system and should be discouraged among women; or at least, significant rest periods should be offered (Strange 2000; Bullough and Voght 2012). According to physicians, menstruation also predisposed women to nervous and hysterical complaints, and paradoxically, the cessation of menstruation—menopause—was also a period of mental derangement (Strange 2000). Under this belief system, women could never escape the tyranny of menstruation at any point in their adult lives. Finally, menstrual blood loss was linked to the illness chlorosis, a disease we would now call anemia—centering the role of menstruation as a pathological process that drains women of their vitality (Wade 1872).

It is no historical accident that menstruation was seen as a reason to prescribe rest. The anthropologists Lock and Nguyen (2018) note that medicalization of women's reproductive life spans is a form of social control. Through the demonization of menstruation, physicians were able to cast women as less intellectually capable, more prone to mental issues, and weaker, robbed of their physical capabilities through menstrual blood loss. The advice to rest is a reason to enforce social withdrawal on women, keeping them from social functions for a significant part of each month. Therefore, these beliefs were not just taboo, but justification for treating an entire sex as inferior—a form of gender essentialism designed to keep women out of spaces of power (such as education, professional work, or voting). Unless, of course, they were poor or working class women, who were expected to complete their work despite the "risk" to their health (Brumberg 1982).

As (White) women entered the medical world as physicians, they began to push back on these menstrual narratives. One of the first women physicians in the United States, Dr Mary Jacobi, authored a prize-winning essay responding to the widely held medical belief that activity during menstruation could make women infertile. In "The Question of Rest For Women during Menstruation," she argues that menstruation is no more of a pathological process than digesting a meal. She also argues against a depleting role of menstruation on blood composition, quoting the physician Paget: "There is no fixed standard to the composition of the blood. From birth onwards the blood and tissues of each creature are adapted to one another, and the maintenance of health depends on the maintenance of their mutual reactions" (Paget 1854 quoted by Jacobi 1877 p. 165). Jacobi partially bases her argument on a concept of physiological homeostasis, although she does not name it as such. She effectively used biological concepts to argue that women were not being constrained by their biology, but instead were marked by the social constraints put upon them based on essentialist biological thinking. While she was successful in convincing the Boylston prize administrators at Harvard that rest was not needed for menstruation, she was less able to persuade the medical establishment that menstruation was a normal function of women's bodies.

Even now, physiology textbooks write about menstruation as if it were a failure to achieve pregnancy, indicating that their views of menstruation have only somewhat improved (Martin 2001). The belief that menstruation is a biological sign of feminine weakness and mental instability persists in popular culture. *Carrie* is one such example, and it is no coincidence that it is a horror movie, providing insight into society's fear of women's procreative capacities. Is it any wonder that women's iron status is also the victim of their menstrual periods?

As older diagnoses such as chlorosis gave way to newer medical concepts such as anemia, the bad blood between iron status and menstrual periods was accepted with little resistance. The biomedical literature that blames women's iron status on her menstruation is vast (Hallberg and Rossander-Hulten 1991). Without knowing the history of the menstrual period in biomedicine, it seems like an easy leap—women menstruate, losing iron-rich blood in the process, and women are also likely to have low iron and/or anemia. The cause seems obvious. Biomedicine has persisted in doing science and medicine with this belief.

The irony of the situation is that iron deficiency anemia in premenopausal women may not be caused by menstruation at all. When men go to the doctor with iron deficiency anemia, they are sent to a gastroenterologist to search for gastrointestinal bleeding. Iron deficiency is a common cause of gastroenterology referral for men. However, this is not the case for women.

For a long time, the rate of gastrointestinal lesions was unknown in premenopausal women simply because they were being diagnosed with their own periods (Annibale et al. 2003). In a study of premenopausal women with iron deficiency anemia, researchers found that 30% had clinically significant gastrointestinal lesions, mostly in the upper gastrointestinal tract, that could cause bleeding (Carter et al. 2008). Another study of premenopausal women with iron deficiency anemia found that about 60% had gastrointestinal lesions, again, mostly in the upper gastrointestinal tract (Annibale et al. 2003). The rate of lesions did not differ between those with regular or heavy menstrual flow. Given the high rate of gastrointestinal lesions, the researchers argued that women with iron deficiency anemia should be investigated for upper gastrointestinal issues, regardless of their menstrual flow. Yet another study of premenopausal women found that about 20% had lesions that could be causing their iron deficiency anemia (Green and Rockey 2004). and another found that among 19 women whose iron deficiency anemia had been diagnosed as caused by their period, 18 of them actually had clinically significant gastrointestinal lesions (Kepczyk et al. 1999).

Despite this evidence, current recommendations of the British Society of Gastroenterology suggest that premenopausal women with iron deficiency anemia should be treated with iron supplementation without further investigation, while men and postmenopausal women with iron deficiency anemia should be investigated for gastrointestinal lesions (Snook et al. 2021). That's right, women who may have serious gastrointestinal issues are routinely undertreated because their periods were to blame, sent away with no further investigation and a recommendation to take an iron supplement. On a brighter note, the American Gastroenterological Association has recently conditionally recommended upper gastrointestinal testing for premenopausal women who are iron deficient, meaning that gastrointestinal causes are no longer immediately ruled out in premenopausal women (Ko et al. 2020).

Beyond upper gastrointestinal lesions, there are even more causes of gastrointestinal bleeding that may contribute to iron deficiency and/or anemia. Gastrointestinal parasites can cause at least mild bleeding, depending on the intensity of the infection. Inflammation may also contribute to reduced absorption of iron (Verma and Cherayil 2017). And truly pathological menstrual bleeding, menorrhagia, has multiple serious causes, deserving more attention than it may get in medical settings. Menorrhagia is reported to be underreported and undertreated in medical settings (Bauman et al. 2020). These misdiagnoses and medical neglect are a direct consequence of the belief that menstruation causes weakness in women. When reproductive-aged women are simply told to take an iron supplement because menstruation renders them anemic, it hides the fact that there may be other, more serious causes of low iron that should be addressed.

Historically, then, society (via biomedicine) has told women their bodies are pathological, incapable of maintaining appropriate iron stores. What if this is not true? What if society creates women who are starved of iron, rather than being starved by their own "natural" bodies? If we can alter our view of menstrual periods as an evolved feature of female biology rather than a pathology, we can begin to develop evolutionary hypotheses for the maintenance of iron homeostasis in the face of menstrual blood loss. What evolved explanations do we have for low iron stores in reproductive aged women that do not rely on "pathological" menstrual blood loss?

Evolution, menstruation, and iron

The problem that evolutionary medicine must confront is that women need iron to reproduce and yet losing iron-rich blood monthly is a normal part of reproduction. Why would evolution favor a reproductive mechanism that depletes a woman of a limited resource like iron, risking her health and her ability to survive and reproduce?

In Chapter 3, I wrote about that long-ago reviewer who pointed out to me that Ariaal women who had returned to having menstrual periods after birth actually had higher hemoglobin than women who didn't, counter to biomedical expectations. Kate Clancy, on her blog *Context and Variation*, picked up this finding and elaborated on it—she saw this as evidence that higher hemoglobin, and by extension iron, was a marker of her body's ability to carry a pregnancy again—her fecundity (Clancy 2011). But this stance makes menstrual periods even more confusing—if menstrual periods happen when women are fecund, but they lose precious iron through menstrual bleeding, how do they maintain their fecundity? How can we reconcile this apparently pathological loss of iron with women's evolved reproductive abilities?

Here, I am going to propose three evolutionary hypotheses for women's lower iron levels. Developing research questions from theory is an important anthropological skill, as is finding ways to test them. This exercise will demonstrate the logic of creating an evolutionary hypothesis and how to evaluate it, using available information in the scientific literature.

Hypothesis 1

The first evolutionary hypothesis is a mismatch hypothesis. In mismatch hypotheses, humans' evolved biologies are mismatched to modern environmental and social contexts. In the case of menstrual periods, women probably have many more of them now than they had over much of human prehistory. In many societies today, the use of birth control technology and reduced family sizes mean that women spend less of their time pregnant. And for some women, greater reliance on infant formulas mean that women spend less of their time lactating. When women are pregnant, they don't menstruate. Lactation also suppresses ovarian function—which helps space women's births farther apart—meaning that they won't menstruate for a while after birth. In societies where women don't regulate their fertility except via lactational amenorrhea, they don't menstruate as much over their lifespan. One estimate puts the lifetime number of menstrual periods in societies like this at around 100, while women in the United States have about 300–400 menstrual periods over their lifespan (Strassmann 1997). In an evolutionary mismatch framework, we would hypothesize that women have low iron because they menstruate more in modern, fertility-controlled contexts compared to contexts that are like the fertility patterns seen over the course of human evolution, where births were spaced apart via lactation and women had more pregnancies over their reproductive lifespan.

The downside of this hypothesis is that pregnancy, and repeated pregnancy, depletes more iron than menstruation ever could. It's estimated that a typical menstrual period causes about 16 mg iron loss via menstrual blood, meaning that menstrual iron losses equal about 0.4–0.5% of total body iron (Miller 2016). On the other hand, a woman permanently loses 840 mg iron across a typical pregnancy (Institute of Medicine (US) Committee on Nutritional Status during Pregnancy and Lactation 1990). Menstruating over 34 months (the length of a typical pregnancy plus two full years of lactation-induced amenorrhea) results in a loss of 544 mg, assuming there is no increase in absorption to compensate. This amount is significantly less than the iron loss of a typical pregnancy. Given the nature of iron homeostasis, it is likely to be easier to recover tiny amounts of iron monthly, when it is lost to menstruation, than repleting one large loss during the postpartum. In this case, we can reason that the mismatch hypothesis is not supported by the current data, and we can conclude that women's low iron status is not due to an evolved mismatch in the number of menstrual periods in a modern women's lifetime.

Hypothesis 2

The second evolutionary hypothesis is that women's lower iron set point evolved as a side effect—a spandrel—of other, adaptive functions related to women's reproduction. The term spandrel was coopted into biology by the evolutionary biologists Stephen J. Gould and Richard Lewontin. Comparing organisms' phenotypic characteristics to the brightly painted spaces between supporting arches in cathedrals, they noted that characters that appear to "stand-out" in a biological organism might not be adaptive, but instead could exist as a byproduct of another evolved structure. Applied to iron homeostasis in women, a spandrel hypothesis would predict that lower iron evolved as a side effect of menstrual blood loss that evolved for another reason. "Some other reason" could be anything, but the first place we should look if we want to pin women's iron status on menstrual blood loss is the evolution of the endometrium, the lining of the uterus.

Not all primates menstruate. Humans have what scientists call "copious menstruation," meaning that we don't just visibly menstruate, we menstruate a lot. By contrast, some primates have no visible menstruation at all. They do shed their endometrium at the end of their cycle, but they resorb the endometrium and blood. Other primates have just slightly visible menses. Primates also vary in the lengths of their cycle, in case you were curious, but that is a tale for another book!

When you look at primate menstruation across their evolutionary relationships, there is a general trend of more visible menstruation among catarrhines (old world monkeys and apes) and less visible or absent menstruation among strepsirrhines (lemurs and lorises), tarsiers, and platyrrhines (new world monkeys). There are several evolutionary explanations for why humans menstruate to a significant degree that incorporate the idea of iron loss into their explanations.

The first evolutionary explanation for menstruation sees menstruation as a cleansing process, allowing menstrual blood to cleanse the body of dreadful things: from a compound called "menotoxin"—presumed to be released from women's sweat and toxic to male infants—to spermborne pathogens. The flushing of spermborne pathogens, proposed by self-taught evolutionary biologist Margie Profit, was presumed to be adaptive in the face of tremendous cost—the risk of iron deficiency. According to Profit, menstruation is costly, wasting valuable resources—including iron—that could otherwise be used for reproduction. Therefore, copious menstruation could only have evolved to protect against harmful infection, with the harm of blood loss being less than the harm of infection. Her idea was decisively debunked by Beverly Strassmann, an anthropologist who pointed out that, among other things, the iron in menstrual fluid would *promote* bacterial growth, rather than cleanse it (Strassmann 1996).

Strassmann, instead, proposed that the endometrium was an energy problem—that human endometriums are too costly to maintain throughout an entire month and that it is more cost-effective to re-grow and shed the endometrium each month if there is no pregnancy. Therefore, menstruation evolved the effect of preserving the energy needed to be fecund each month. Strassmann also argues against Profit's assertion that iron loss is a huge cost of menstruation, noting that for much of human evolution (without contraception), women probably did not menstruate much, instead spending most of their reproductive lifespans either pregnant or in lactational amenorrhea. She states, "Menstruation must have been similarly rare over human evolutionary history, and the iron lost through menses trivial relative to dietary intake. Anemia would have been especially unlikely among hunter-gatherers whose diets contained significant quantities of meat (Strassmann 1996 p. 202).

Finn (1994) proposes a different hypothesis for menstruation. He notes that menstruation is the natural endpoint of the endometrium in humans—after being primed with reproductive hormones, endometrial tissue differentiates to a "point of no return" in the latter half of the menstrual cycle that cannot be reversed and must be shed so that the endometrium can proliferate during the next menstrual cycle. This can help explain why humans menstruate while other mammals, like mice, do not—women's endometrial differentiation is triggered by hormones, while in mice the endometrium differentiates only when a blastocyst (fertilized clump of cells) is present, meaning their endometrium changes only when a pregnancy is imminent. Finn hypothesizes that terminal endometrial differentiation in humans is necessary because of the invasiveness of the human placental trophoblast (the part of the placental that embeds into the uterus) and that menstruation was a byproduct of the human implantation process. He counters Profit's assertion that menstruation is wasteful by noting: "However, the fact that women have survived this biological loss is evidence that, from the point of view of evolution, there is no great pressure for its removal. Waste is in fact a feature of biological life . . . . Women probably lose much more valuable biological material in their faeces than in their menstrual fluid (Finn 1994 p. 1203)." While Finn does not explicitly specify iron as a valuable "biological material," it is likely that iron is one of the biological materials that he meant, especially since Profit argued that iron loss from menstruation was costly.

Clancy, in examining Strassmann and Finn's hypotheses for the evolution of menstruation, notes that Finn's hypothesis also allows for the endometrium to respond to the ecological conditions of the mother (Clancy 2009). She notes that the ovarian hormones that differentiate the endometrium have been shown to be sensitive to energetic status and expenditure, diet, nutritional status, and developmental conditions, with levels of these hormones ranging widely between and within populations. Further, she found that endometrial thickness itself is sensitive to nutritional status—specifically, iron status. She found that the thicker the endometrium a woman has, the higher her hemoglobin level is, a direct contrast to the biomedical belief that menstruation causes iron loss. Instead, it appears that the women with sufficient somatic resources to maintain a thicker endometrium do so, regardless of the possibility that it will be lost due to menstruation. Ultimately, she concludes that normal menstrual periods do not cause anemia (Clancy 2011; Clancy, Nenko, and Jasienska 2006).

These hypotheses for the evolution of menstruation do acknowledge that there is some loss of iron (or "biological material") due to menstruation but tend to view it as a small cost of doing the business of human reproduction. For what it's worth, iron status is not necessarily the main test for these hypotheses but is instead presented as a potential cost of menstruation without quantifying that cost. This is why, under these explanations, women's lower iron set point is a spandrel. It's a fancy, distracting characteristic that does not reflect a direct adaptation itself, but instead is a tiny side effect of these other, adaptive changes to the efficiency of endometrial maintenance during the menstrual cycle.

In fact, there is quite a bit of uncertainty around the true nature of iron loss as a cost during menstruation: How much iron is actually in menstrual fluid? How much is reabsorbed before being shed with endometrial tissue? Does the body have homeostatic mechanisms to absorb more dietary iron while menstruating? Unfortunately, testing evolutionary hypotheses for menstruation cannot completely answer our question about the evolution of low iron in women, especially since iron is treated as a spandrel. Evolutionary hypotheses that discuss the costs or benefits of low iron status explicitly are better able to answer why women's iron physiologies have lower set points.

Hypothesis 3

The third hypothesis that could explain women's lower iron set point is that iron physiology evolved to be that way on purpose—that low iron in women is adaptive. This seems like a shocking, out-of-touch hypothesis that is not in line with a century of received biomedical wisdom, nor of evolutionary thinking that casts low iron as a small cost of reproduction. Even so, let's consider the rationale and evidence.

There are two parts to consider in this hypothesis. First, there should be a reason that lower iron is adaptive—a way that the feature (low iron) would improve survival and reproduction for women. Second, there should be an evolved mechanism that keeps iron levels lower, something that impacts the physiology of iron homeostasis. My description of iron homeostasis during pregnancy in Chapter 3 lends us a clue to the possible adaptive nature of low iron. Remember, hepcidin is high in the first trimester, preventing women from absorbing iron. Researchers hypothesize that this mechanism exists to protect the developing embryo from the harmful, DNA-damaging effects of iron (Weinberg 2010). This mechanism won't work, however, if women's iron levels are already high to begin with. What if reproductive-aged women's lower iron set point exists to protect *potential* embryos from harm? That means that there could be a mechanism that keeps women's iron levels just a little bit lower, in case a menstrual cycle results in a pregnancy. In this section, I'm going to start with the second point—what mechanism keeps iron low in women?

It is easy to point fingers at pregnancy or menstruation as a cause of women's low iron. There are multiple studies that have quantified the amount of blood lost during menstruation—an average of 16 mg of iron lost per period, which seems like an amount that could overwhelm a woman's system. This makes it appear as if women's physiology is desperately racing to recover iron loss each menstrual cycle. However, despite the considerable amount of science that has quantified menstrual blood loss, few studies have explored how iron is recovered after a menstrual period—a shocking omission considering that iron metabolism is governed by the principle of homeostasis.

A recent study has provided new insight into iron homeostasis surrounding menstruation. Dubbed the HEPMEN (HEPcidin during MENses) study, this project looked carefully at serum iron and hepcidin over the menstrual cycle in a sample of women, beginning with the first day of menstruation (Angeli et al. 2016). They modeled serum iron level and hepcidin level separately and together, demonstrating that the two had intricately linked dynamics. Their results show an initial decrease in serum iron over the first 5 or so days of the menstrual cycle (see Figure 4.2), which coincides with menstruation. This shows that iron is depleted during menstruation. The hepcidin

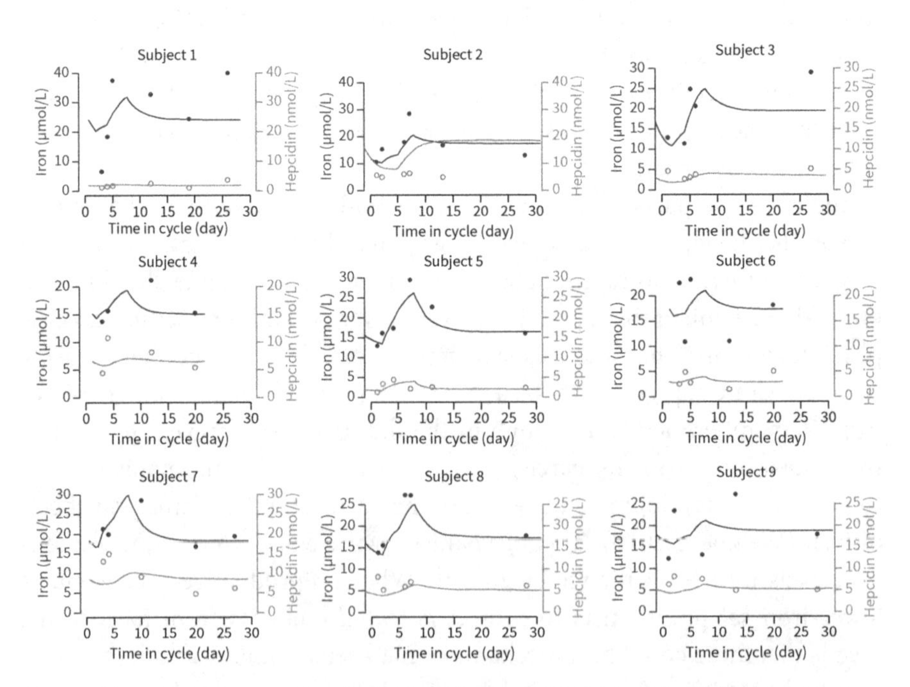

Figure 4.2. Levels of serum iron (red, closed symbols) and hepcidin (blue, open circles): Individual fits for the first nine subjects in the dataset. The symbols on the individual fits represent individual measured concentrations, and the solid lines represent model predictions.

Source: Reprinted from Angeli et al. (2016) with permission from Springer Nature.

level decreases slightly after the serum iron level decreases, meaning that the body is primed to absorb more dietary iron. From there, serum iron levels rise dramatically and overshoot their initial levels by day 7. Hepcidin levels rise soon after, slowing the absorption of dietary iron. By day 10, both serum iron levels and hepcidin levels have normalized to a set point. If you want to illustrate physiological homeostasis—the tendency of physiology to reach a stable equilibrium—it does not get better than this figure. Menstruation seems to destabilize body iron's set point, but within days hepcidin helps iron levels compensate—even overcompensate—and then eventually stabilize to a consistent level. This means that the amount of blood loss associated with a typical menstrual period does not likely permanently reduce women's iron set points. Indeed, the HEPMEN study demonstrates that it is possible for women to achieve even higher levels of serum iron than they do. What could be the reason for that?

The HEPMEN study also provides insight into what characteristics create individual variation in hepcidin and iron levels, including the maintenance of a physiological iron set point. The first is the women's Higham score, a pictorial chart used to estimate the amount of menstrual bleeding occurring in the cycle. A higher score was associated with a greater spike of hepcidin during menstruation, meaning that hepcidin was compensating for a greater amount of iron lost due to menstrual blood. Interestingly, Higham score was not linearly related to the amount of serum iron in the body, with one interpretation being that losing more menstrual blood did not significantly impact homeostatic iron levels. Ferritin level, a marker of iron storage, was associated with less reduction of hepcidin and smaller rebound of serum iron after menses, which means the fluctuations of hepcidin and serum iron were less pronounced throughout the cycle. Having more stored iron, then, meant that iron absorption mechanisms were not deployed as strongly compared to women with less stored iron. This is an expected dynamic in a system designed to reach homeostasis—the women with more stored iron need less iron absorption to reach equilibrium.

The other factors predicting variation in the HEPMEN study are also interesting. Body Mass Index (BMI) was a significant covariate, with women with a higher BMI unable to muster as big of a serum iron rebound after menses compared to women with lower BMI. The HEPMEN study authors suggest that this may be due to the relationship between inflammation (which is higher in people with higher amounts of adiposity) and iron homeostasis. I'll talk about this more in Chapter 5, but inflammation is a powerful signal to the body that free iron should not be released from storage! Therefore, it makes sense that serum iron does not significantly rebound if inflammation is high. Since they didn't measure inflammation directly in their study, however, their conclusion is still a hypothesis. Interestingly, height also mediated the cycling of hepcidin. Taller women had a greater release of hepcidin at the end of menses, showing a much greater rebound effect than shorter women. No explanation of this finding was given by the authors, other than to say that it was a strong and surprising effect, and that body size beyond weight or BMI should be taken into consideration when modeling iron homeostasis.

The final—and most intriguing—factor is that taking birth control modifies iron levels. This is a well-established finding from other studies—women who take hormonal birth control have consistently higher iron levels and other indicators of iron status (Milman, Kirchhoff, and Jørgensen 1992). They also have less menstrual blood loss. The conclusion of scientists who observed these two facts was that women taking birth control were protected from too much menstrual blood loss, resulting in higher iron levels. That's what the HEPMEN study authors concluded—that menstruation in women without hormonal birth control results in 20% lower serum iron and that this was attributed to the greater amount of iron lost via menstrual blood among these women. Their own results, though, showed that serum iron was not related to menstrual blood loss (Higham score). Instead, Higham score was associated with hepcidin. This means that menstrual blood loss activated hormonal control of iron recovery. Homeostasis worked: Women who menstruated more also activated iron absorption more strongly, which then settled into an iron set point. The fact that birth control pills are associated with a higher iron set point—and it is useful to think of the birth control condition as the "not-normal" condition rather than the normative one—might mean something other than blood loss regulates women's baseline iron levels. Could reproductive hormones be the mechanism keeping iron levels within certain parameters for women?

There is firmly established evidence that at least one reproductive hormone regulates iron status—androgens. This class of reproductive hormones is associated with "maleness"—testosterone is the most well-known androgen—but female reproductive organs also produce low levels of androgens. Androgens are the leading explanation for differences in iron levels between men and women. Androgens are known to facilitate the creation of red blood cells (erythropoiesis). In fact, there is a dose-dependent relationship between testosterone and red blood cell production among men (Coviello et al. 2008; Bachman et al. 2013), a factor that may in fact be associated with greater risk of cerebrovascular disease (Gagnon et al. 1994). Ironically, men with higher testosterone tend to have lower ferritin levels, likely due to the mobilization of iron into red blood cells and/or cross-talk between iron, inflammation, testosterone, and obesity (Chao et al. 2015). Women with polycystic ovarian syndrome (PCOS) also have an excess of androgens compared to most reproductive-aged women. Along with this, women with PCOS also have higher iron stores in the form of ferritin (along with increased inflammation and obesity), putting women more at risk for insulin resistance and diabetes (Escobar-Morreale 2012). This suggests that high iron is a sign of hormonal and metabolic disruption, at least among reproductive aged women.

Menopause, the end of menstrual blood loss, is another data point in the relationship between reproductive hormones and iron status. By menopause, women's progesterone and estrogen levels are extremely low. By contrast, iron stores rise—not to men's levels, but higher than they are in reproductive aged women. Along with higher iron and lower estrogen/progesterone, the post-menopausal phase of life is associated with higher risk of cardiovascular disease and other chronic illnesses. Researchers note that while most research on these health risks after menopause use the withdrawal of estrogen as the explanatory factor, higher iron is also a risk factor for increased chronic illness such as Type II diabetes, metabolic syndrome, and cardiovascular disease (Bozzini et al. 2005; Kobayashi et al. 2018). Higher androgens cannot be the explanation since they are also low postmenopause. Could low estrogen and high iron be directly related to each other?

It's exceedingly difficult to figure out the relationship between female reproductive hormones and iron. For one thing, there are so many moving parts during the menstrual cycle—estrogen and progesterone rise and fall with ovulation and menstruation—that it is difficult to test their effects experimentally. An observational approach like the HEPMEN study might be useful in reproductive aged women, but it can't necessarily account for the loss of estrogen and progesterone at menopause. This makes testing this hypothesis extremely difficult and time consuming. While exploring data for another publication, however, I stumbled on an observational experiment that could possibly be used to evaluate the relationship between hormones and iron. I was looking at the National Health and Nutrition Examination Survey (NHANES), a multi-year sample of Americans that is meant to be a representative sample of the health of the people of the United States. There is an overwhelming amount of data available, including multiple measures of iron status. There was also, to my surprise, information about the use of hormone replacement therapy (HRT) in postmenopausal women. The HRT variable, which was divided into estrogen only, progesterone/estrogen, and progesterone only, allowed me an interesting observational look at the effects of these hormones on iron status in women over the age of 45.

Most postmenopausal women on HRT take an estrogen/progesterone combination. Estrogen, it turns out, promotes endometrial growth and endometrial shedding, while progesterone reduces endometrial shedding. This is a function of these hormones' purposes: progesterone facilitates pregnancy, meaning that it works to keep the endometrial lining in place, while estrogen promotes the growth and maturation of the endometrium during the first part of the menstrual cycle (follicular phase). Without the tempering force of progesterone, the endometrium can grow too much, leading to excessive bleeding. And yes, this can happen to postmenopausal women, too. I was interested to find out that in addition to an HRT variable, the NHANES also had a variable that designated whether a woman had a hysterectomy, that is, whether her uterus had been removed. This would help me control for the possibility of breakthrough bleeding while on HRT, because women without uteruses would not have an endometrium. I ended up with four groups to compare: no HRT/no hysterectomy, no HRT/hysterectomy, HRT/no hysterectomy, and HRT/hysterectomy. I wanted to see how indicators of iron status (ferritin and hemoglobin) were different between the groups. I hypothesized that iron would be lower in women using HRT and that having a hysterectomy would have no difference on iron status, because I thought hormones helped regulate iron homeostasis, not blood loss from the endometrium.

I was partially correct. It turns out, ferritin levels were indeed lower in women who were on HRT. However, they were also lower in women who still had their uterus, meaning that it is possible that iron was lost via endometrial breakthrough bleeding. Since there wasn't a variable indicating whether a woman experienced bleeding while on HRT, it's impossible to be certain. This indicated to me that breakthrough blood loss and reproductive hormones could both play a role in keeping iron levels lower, at least in postmenopausal women. It's not clear how much of this result can be extrapolated to reproductive-aged women, but it looks increasingly like menstrual blood loss alone is not responsible for keeping iron levels low in women. Female reproductive hormones do part of the job, too.

When I wrote that article, I focused on estrogen as the causal factor for the low iron. There is some evidence that estrogen affects iron homeostasis in female mice (Hou et al. 2012), and so I thought that it would be a major factor in the lower iron levels in reproductive aged women. In hindsight, my interpretation was not entirely correct. More research, performed since that paper was published, has established that estrogen inhibits hepcidin expression, meaning that absorption of iron, and thus iron levels, would be greater when estrogen is higher. However, estrogen is probably not acting alone. Progesterone may also effect iron homeostasis. In fact, evidence is growing that progesterone increases hepcidin levels, which would decrease absorption of iron and cause lower iron levels (Li et al. 2016; Barba-Moreno, Alfaro-Magallanes, Calderón, et al. 2020). In reproductive aged women, this would mean that estrogen would increase iron absorption early in the menstrual cycle (helping recover from menstruation) and progesterone would decrease iron absorption later in the menstrual cycle (to protect a potential pregnancy from iron overload). The push-and-pull actions of progesterone and estrogen may create a set point for iron status that is distinct for reproductive aged women. The postmenopausal women, who are not experiencing HRT in the same manner as endogenous hormone secretion in reproductive aged women, may not experience a lower iron set point in quite the same ways as premenopausal women, but it is still enough to affect their physiology. Estrogen and progesterone may also impact inflammatory processes during the menstrual cycle, such as interleukin-6 secretion (IL-6, a pro-inflammatory marker), and this inflammation may also impact iron status at certain points in the menstrual cycle. This suite of physiological changes across the menstrual cycle might also possibly explain the opposing effects of birth control pills and HRT on iron status.

The evidence that there is hormonal control of hepcidin and women's iron status is growing, with an unclear path forward for testing hypotheses in humans. It appears that in some non-human studies estrogen and progesterone impact iron homeostasis, with progesterone increasing hepcidin (which lowers iron absorption) and estrogen lowering hepcidin (which increases iron absorption). A study design such as the HEPMEN study, but with added reproductive hormone measurements, might be able to capture an in vivo picture of how these hormones and inflammatory markers dynamically intersect with iron status across the menstrual cycle. One study that looked at reproductive hormones, IL-6, iron status, and exercise across three phases of the menstrual cycle found that both estrogen and progesterone affect IL-6 levels, but that there were no significant effects of these hormones or IL-6 on hepcidin or iron stores (Barba-Moreno, Alfaro-Magallanes, de Jonge, et al. 2020). However, since this was a sample of only 15 women, more work is needed to capture small effect sizes, if they exist.

So, given the evidence, where does evolutionary hypothesis three stand? There are two pieces needed to support it: (1) an adaptive explanation for a lower iron set point among reproductive aged women and (2) an evolved mechanism that ensures reproductive aged women's iron levels remain low.

For the first piece, there is good evidence for two adaptive explanations for a lower iron set point, discussed in Chapter 3: Preparation for an embryo that faces developmental harm from iron, or protection for the mother against infection, inflammation, thrombosis, and hypertension in pregnancy. In both cases, the adaptations would be protective against harm done by iron, a common theme in the history of iron evolution. Protection from harm seems to be the leading adaptive explanation for women's lower iron set point.

The second piece is less clear. Menstruation and especially pregnancy both seem to lower women's iron stores. In the case of pregnancy, differences between male and female iron levels appear in puberty before most girls could become pregnant, so pregnancy is not the likely cause of pubertal iron dimorphism. With menstruation, evidence suggests that regular iron homeostasis is capable of returning serum iron to their normal levels, especially in those who have sufficient stores of iron in the first place. Since iron stores are only lower during the reproductive years and increase during menopause, this may also point to the possibility of reproductive hormones altering iron homeostasis. Since taking reproductive hormones (birth control and HRT) raises iron levels in reproductive aged women but lowers them in postmenopausal women, the mechanism of action is not clear, and observational studies are needed to figure out the complex dynamics of hormonal changes and iron status in reproductive aged women.

Finally, part of this discussion should be turned on its head: We should be asking why testosterone has the effect of increasing iron stores rather than asking why women's iron is so low. Maybe we should be treating women's iron stores as normal and men's as different—especially since higher iron stores are associated with cardiovascular illness and other negative health effects. The implicit belief that "more is better" when it comes to iron status may not be true, especially when it has been men (who have more) who are seen as normal and women (who have less) are seen as pathological. The inherent sexism in this belief goes hand in hand with the sexism and taboo associated with menstruation—that it is always associated with women's natural weakness. Perhaps it is time to turn that belief on its head. We should be stating that menstruation is a sign of health, and that normal menstrual iron loss leaves no lasting pathology.

Conclusion

Menstruation is a useful boogeyman for biomedicine: it can be blamed for women's iron deficiency and provides an acceptable biomedical explanation for women's weakness. This bias against menstruation is not just confined to biomedicine; rather, I contend that it is so widespread that it has biased whole academic fields away from doing good science, taking for granted that menstruation saps iron without careful hypothesis testing. Even human evolutionary science, which is critical of biomedical conclusions, has been slow to investigate this misconception. In this chapter, I've presented a hypothesis that women's low iron is not due to their menstruation, but instead evolved to protect their bodies and the bodies of their offspring from harm during reproduction. I've also proposed a possible mechanism that implicates female reproductive hormones, one that is grounded in the physiological systems that underly homeostasis. Unfortunately, the dynamic nature of homeostasis means that the mechanism underlying my proposed evolutionary hypothesis will not be easy to test. However, science is long overdue for a critical, complex look at menstruation, reproductive hormones, and iron homeostasis. The reasons that it has not done so—the structural sexism in science—are as clear as day once you trace their origins to their historical source. Dr Jacobi would be shaking her head in disbelief.

This brings up a larger issue for scholars: we should be asking ourselves why our first instinct was to blame menstrual periods for women's low iron without a more careful consideration of the homeostatic mechanisms of iron status. There are multiple examples of how sexism can impact how scientists ask and approach research questions—the old "Man the Hunter" trope in human evolution (Slocum 1975), the idea that there are "male" and "female" brains (Fine 2010), or even that eggs passively wait to be penetrated by sperm (Martin 1991)—and women's iron status may be yet another example. Why has it taken so long to ask evolutionary questions about women's iron homeostasis? Why haven't there been studies of the effects of female reproductive hormones on iron status? If you've already decided menstruation is the explanatory pathology, of course, then no more study is needed. And women, despite being half of the population, are never treated as the default in human biomedical studies (Mazure and Jones 2015), a fact that has put women's health at grave risk. This should be a reminder to all scientists that questions are never settled, fresh perspectives are important, and that *who* is doing the science is just as important as *what* the science is. All fields need a Dr Jacobi—and more—to remind us that established facts can always be subjected to better science.

Curiously, this corner of our mosaic is also incomplete. It looks as though there should be other people in the image. There is room for a whole choir, but instead there are only a few disjointed singers. Disappointed, we turn to the opposite end of the mosaic. While we have mostly focused on the part of the image that shows iron leaving the body, this side of the mosaic shows iron entering the body. We have been focusing on only one part of the story, the evolutionary narrative. Maybe this side, the intake of iron, will flesh out this tale. We peer closer, spotting a woman eating a meal with her family.

5

Gut Feeling

The Gut as a Nexus between Microbiome, Diet, and Iron

In another corner of our mosaic, we see a montage of a woman's day. She rises in the morning and makes breakfast. She prefers hot, milky tea, but she prepares porridge for her growing children. Her family's water sources are far away from home, so she and her daughters spend considerable time collecting and bringing water back. She leaves her home to go to work, since she is a major wage-earner for her family. She is a domestic laborer, doing housework and watching children for a wealthier household. For lunch, she eats what is left over from her charges' lunch, usually the local staple food and occasionally a piece of fruit. When she leaves after a 12-hour workday, she has a long commute: She can only afford to live an hour away from her employer. When she finally gets home, she cooks a typical meal of staple food, vegetables, and meat. She takes a helping of the staple and saves the rest for her husband and children. The mosaic shows the weariness etched into the woman's face.

Embodiment

The phrase "your body is a temple" is widespread among the health-and-wellness crowd in the United States. This group, largely a White upper/middle class aspirational industry geared toward women—think Gwyneth Paltrow and her multimillion-dollar GOOP company—spends substantial amounts of time and money on wellness and "self-care." This style of "wellness" reflects the belief that the body has an inside and an outside, and that the individual person can control what gets inside. "Inside" is a pure, almost sacred space, untouched by the polluting forces of the outside world. If you've been polluted, though, you still have hope. You can purify your body if you can afford the right things. Examples of purifying substances include certain foods, supplements, powders, aphrodisiacs, lubricants, essential oils, tinctures, "fasting" and "detox" products, and "beauty ingestibles;" while polluting forces include the wrong kind of food, "toxins," and vaccination

Thicker Than Water. Elizabeth M. Miller, Oxford University Press. © Oxford University Press 2023.
DOI: 10.1093/oso/9780197665718.003.0005

(while GOOP does not recommend against vaccination, vaccine hesitancy is widespread in the alternative medicine sphere). Above all, the ability—and obligation—to control what goes in and out of the body is a key tenet of wellness. Maintaining the purity of the body's borders is paramount for good health.

This belief is in direct contrast to the anthropological concept of embodiment. Embodiment acknowledges that the world, especially the social world, "gets under the skin" in a whole host of ways. Interpersonal experiences, the culture-bound nature of what we consume, the social structures that shape how we live our lives, and more, all play a role in creating our body's biology. In this framing, the body's borders are porous and in continuity with the rest of the world—there is less distinction between "outside" and "inside" the body than might be readily apparent. Margaret Lock proposed the concept of local biologies to explain how the confluence of social structure, cultural practices, and ecologies become entangled with the body's biological processes to create unique health circumstances in specific localities (Lock 1994). This might explain, say, how agricultural practices and ineffectual healthcare structures in rural areas make a West African population vulnerable to malaria. Or it might explain how living in a particular postal code is a major determinant of health. Studying embodiment relies on a thorough examination of social circumstances, environmental exposures, and human biology.

Embodiment is used to explain health disparities. Health disparities are different health outcomes for different groups based on the group's social position. Health disparities due to poverty are widespread throughout the world, including developed economies like the United States. Those living in impoverished conditions tend to face poorer health, but the type of poor health they experience is typically local: for example, the health disparities in tropical areas may tend to have more infectious disease, while poverty in wealthier nations may have a greater prevalence of chronic disease. Thus, the illness experienced due to poverty is both global and local.

While poverty is the biggest global source of health disparities, national and local conditions can force other types of social disparities to the fore, such as disparities by race, ethnicity, gender, sexuality, nationality, religion, and more. For example, in the United States there are significant health disparities based on the social category of race. The country's long history of enslavement of Black people and subsequent policies meant to keep Black populations disenfranchised manifests as a shocking array of poorer health outcomes for Black people in the United States. Intersections with other social identities can also make health disparities worse, so for example, those who are Black and poor have worse health outcomes than those who are Black and well-off.

The United States also has a history of exploitative and violent gynecological and pediatric practice, reflecting the status of women as second-class citizens for much of its history (Owens 2017). The identity of "women" and the identity of "Black" intersect to mean that Black women have exceptionally poor health outcomes in the United States.

A major downside of embodiment theories is the "black box" nature of how the social world impacts the body and its processes. It is easy to see the relationship between a broad social force and health outcomes; it is much harder to see the pathways by which these social phenomena manifest within the body. Since embodiment, as a theory, is primarily used by social scientists, the focus on the social processes makes sense. But it can be frustrating to biological anthropologists who are used to quantifying bodily measurements. In general, pathways of embodiment are thought to be neuropsychological: The experience of events is taken in by the central nervous system and processed by the brain, which signals that the situation is stressful. The cascade of events following a stress response can alter how multiple systems of the body operate, particularly if the stressful experiences are chronic.

However, there are other routes through which embodiment can occur—inhalation of pollution, the absorption of phthalates through the skin, and the consumption of food and water are all means of entry to the body. Diet, while not always considered a pathway of embodiment, is absolutely a mechanism through which the social world gets under the skin, substantially altering its physiology. Since many of the mechanisms of diet are already "known," this is where biological anthropologists can be helpful. We can link the social world of food and diet to known mechanisms of digestion and absorption. Further, we can incorporate new understandings of our complex biological makeup—such as the microbiome—into the pathways of embodiment that we already know. Biological anthropologists are therefore well positioned to crack open the "black box" of the body as it is conceived in social theory.

Biological anthropologists studying embodiment face another problem: The operationalization of the social. It can be difficult to take the contextualized experiences that cultural anthropologists relay via rich ethnography and render them into a series of numbers so that they can be statistically compared to a biomarker (which is also a snapshot of a bigger biological process!). Further, just as there are proximate and ultimate levels to biology, so too there are levels to social organization. Some reflect broader aspects of social organization, such as political economy, globalization, ecology, and social institutions. These affect individuals and groups and are the social structures that shape an individual's social positioning and possibilities. Others are more "proximate" to the everyday experience of the individual, such as kinship and other forms of social organization, beliefs and systems of meaning, dietary practices, and other aspects of culture. Of course, these social factors do not exist in a rigid hierarchy, and there are constant self-reinforcing continuities between them. The relationship between these levels—how structure integrates with practice, for example, or how power is enacted to discipline individuals into accepting a particular form of "normal"—is a topic of concern to the social sciences, and just a bit beyond the scope of this chapter. In this chapter, I'll focus on "closer" aspects of social life, discussing the mechanisms that signify the pathway of embodiment of iron—in this case, the pathway of the dietary absorption of iron, with the nexus of embodiment being the gut and its inhabitants. In Chapters 6 and 7 I will focus on bigger social structures like race, poverty, and gender and how they manifest into iron deficiency (Figure 5.1).

Embodiment as a framework has much to offer the human evolutionary sciences. The evolutionary medicine concept of mismatch holds that humans' evolved bodies are "out of place" in modern environments, and that many illnesses exist because our bodies cannot handle modern diets, sedentism,

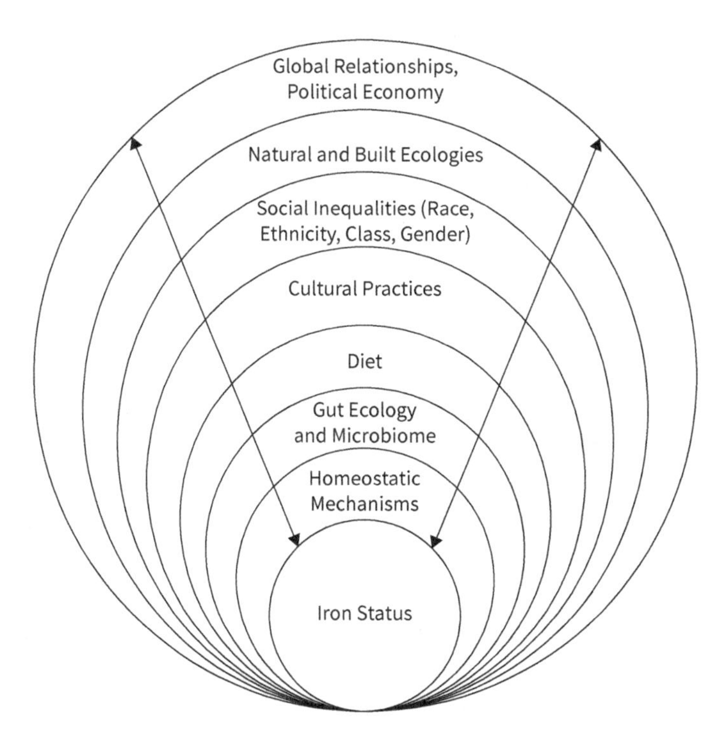

Figure 5.1. A rough conceptual model of the pathways of embodiment. These pathways may be bidirectional and entangled in a multitude of ways that are difficult to depict visually.

and indoor lifestyles. However, all of these "modern" environments have come about due to human social behavior, meaning that in-depth use of social theories and frameworks are a complementary, and even necessary, part of studying evolutionary medicine. Embodiment offers the ability to contextualize how modern environments are created via social action and to hypothesize how these environments become part of the human phenotype. This is one way the simple evolutionary medicine concept of "modern environments" can become more sophisticated, allowing for improved understanding of the relationship between our evolved bodies, social worlds, and illness.

Medical anthropologists use the term embodiment to discuss how the social world "gets under the skin" of people to create illness. When it comes to iron deficiency, the gut is the primary way that the iron outside becomes the iron inside. In this chapter, I frame the gut as a nexus for the embodiment of iron status, discussing the links between the gut, microbiome, diet, craving, and nausea. First, I'll talk about the basics of iron absorption and movement throughout the body, discussing the effects of "modern environments" on iron physiology. Next, I will discuss the gendered nature of foods in many societies—with women's food often coming out on the bottom, iron-wise. Next, I will analyze edible cravings, particularly during pregnancy. Craving dirt, clay, and chalk, a phenomenon known as geophagy, is associated with iron deficiency, a relationship that puzzles scientists. Finally, I will explore how humans are not the only species absorbing iron in our guts—a host of iron-metabolizing bacteria also consume dietary iron with consequences for human well-being. This microbiome may also impact craving, making it a bidirectional actor in the embodiment of iron. Above all, I will emphasize the cultural contexts of how we consume, absorb, and discard iron via the gut.

Getting iron under the skin: The gut as a nexus of embodiment

As a new graduate student in anthropology (almost two decades ago!) I was casting about the literature to find a potential project. I needed a great idea, one that I could write about for a fellowship application. I really needed the money, and I wanted to prove to myself and others that I was a young force to be reckoned with in the field. I somehow stumbled upon an issue of *Science* that permanently changed the way I thought about the human body. Entitled *Gut: The Inner Tube of Life*, this issue focused on the evolution, structure, microbial flora, metabolic functions, and the immune system of the gut—a satisfying mix of evolution, development, and physiology that was incredibly appealing to a young human biologist. Despite the weird title, I was hooked—it was amazing how much was going on in the gut, how it received *stuff* from the outside environment and in turn fed microbes, fed the body, and kept the immune system on its toes. While I ultimately decided I couldn't directly measure gut function easily in the field at the time, my dissertation work on human milk immunity was impacted by the fact that the infant gut served as an interface between a mom and her baby, one that impacted infant growth and immunity into adulthood. Future work has confirmed that the mother's milk and baby's immune system are inextricably linked, meaning that my old hypotheses were not far off base. And I also won the fellowship, teaching me that it was important to think big even if your projects must be smaller.

As I have done more work on human milk, the microbiome, and immune function, it has become increasingly clear that the gut is a nexus—a connection—between the outside world and the inner workings of the body. It is one of the few ways that material from the outside world can be absorbed inside the body: digestion. It has its own vast immune system, maintaining a special porous barrier between the outside and the inside. It is also a gathering place for a group of microbes—mostly helpful—to snack on digested substances and provide their own help (or harm) to the gut. This bacterial gallimaufry is in regular contact with the gut, creating complex communities that facilitate digestion and mediate human immune responses. This makes the gut a perfect pathway of iron embodiment.

Diet is the only mechanism for introducing new iron into the body (not counting medical interventions like transfusions). There are two types of dietary iron: heme iron and non-heme iron. Heme iron bound to proteins like myoglobin (muscle) and hemoglobin (blood) and is found in animal dietary sources. Non-heme iron is bound to non-heme proteins and are typically found in plant foods. While both forms of iron can be absorbed by the gut, relatively little iron makes it into the body from the diet. Heme iron is more easily absorbed than non-heme iron, meaning about 15–35% of the iron in a heme food source is absorbed compared to about 2–20% of non-heme iron. These numbers are variable because the gut tightly controls how much iron is absorbed via hepcidin, and if the body already has sufficient iron then less iron will be absorbed from the diet (see Chapter 2). Many other dietary substances also alter how much iron is absorbed: Vitamin C enhances the absorption of iron, especially non-heme iron; while phytates and polyphenols (present in beans, legumes, tea, and grains), as well as certain animal and plant proteins (such as casein, whey, or soy proteins), oxalic acid (found in dark leafy greens, beans, and nuts), and dietary calcium inhibit iron absorption. The push and pull of dietary iron inhibitors and enhancers seem very clinical, but in practice, the choice and consumption of foods is dictated by culture—meaning that the causes of embodied iron status are beholden to social behaviors.

In addition, iron supplements vary in their ability to be absorbed. Supplemental iron is usually found in various ferrous and ferric mineral salt formulations, such as ferrous sulfate or ferric citrate. Absorption of these mineral salts depends on hepcidin levels and iron status (of course), as well as the general bioavailability of iron within each salt. The ferrous ion of iron is more easily absorbed because this is the form that crosses through the intestine. Ferric iron must be converted in the acidic stomach to be absorbed, and that additional step means that ferric iron is less bioavailable.

Given the variety of ways iron can be helped or hindered by dietary factors, it should be no surprise that absorption itself is a complex dance at the border of the small intestine. Most iron is absorbed in the first part of the small intestine, right after leaving the stomach. The acidic nature of that part of the small intestine helps convert insoluble ferric compounds into soluble ferrous compounds. It is here where iron absorption enhancers or inhibitors have an effect—these can alter the pH of digested material or form compounds that alter the ability of iron to be absorbed. Once iron is in ferrous form, it is transported into enterocytes, or cells of the intestinal lining, where it is stored as ferritin or further passed through the intestine and into circulation via ferroportin (Chapter 2). Hepcidin, of course, helps regulate ferroportin, so how much iron is passed via ferroportin is dependent on iron status. Given how many factors alter iron absorption, it is no surprise that only a fraction of it makes it into the body.

Since most dietary iron isn't absorbed, it is either passed out of the gastrointestinal tract via stool or is used by the millions of resident microbes cohabitating in the gut. This microbiome is home to friendly organisms with a variety of effects on human metabolism and immune function. As I'll discuss later in the chapter, iron feeds certain microbial residents, altering host physiology and the habitat of the gut. While this is a new area of research for iron metabolism, a small but growing body of evidence suggests that there is a complicated relationship between humans, our microbiome, and the iron we consume—one that might even alter our behavior.

Infection is also a major modifier of iron status, changing how iron gets into the body and what happens to it when it is in there. Iron withholding is a physiological response to infection in which iron is mobilized from circulating macrophages and moved to iron storage proteins where infectious microbes cannot access it (see Chapter 2). Iron withholding also changes the physiology of gut iron absorption, in that it reduces the amount of iron absorbed from food (Schmidt 2015). Iron withholding is not triggered by infection, exactly, but is instead triggered by inflammation, which is the body's response to infection. However, inflammation can be present in the body without infection. Chronic illnesses such as autoimmune disease, metabolic dysfunction, or even an excess of adipose tissue can lead to heightened levels of inflammation in the body in the absence of infection. Hepcidin, the iron metabolism hormone, is sensitive to the effects of inflammation and thus mediates iron withholding processes in the gut. When inflammation increases, hepcidin goes up, preventing iron absorption.

This is the physiological nuts-and-bolts of iron absorption. While culture, "modern environments," and food choice can all impact how the body takes in iron and moves it around, human cultural behavior constrains the types and amounts of foods people eat. Cultural beliefs and practices shape who eats what within and between societies. For women, this can mean eating foods that lack sufficient iron for their body's needs.

What we eat

Women can eat anything?

There is a wealth of anthropological information about the culture-bound nature of diets. I was first exposed to this idea when I visited northern Kenya and learned about the concept of gendered food consumption, or the idea that there are foods for men and foods for women. The Ariaal people who participated in my research were settled pastoralists, meaning that they did not move their homes from place to place, but instead lived in one place and did some subsistence farming. They also kept livestock: cattle and small livestock like goats and sheep. Since cattle are expensive and take a long time to grow, they are usually not used as an everyday source of meat. Instead, female cattle are used for milk; and pastoralists will also drink their blood.

Cattle blood is also a "sometimes" food for the Ariaal. It is most frequently used when herding cattle in arid environments for many days. The young men that herd the cattle will drink blood, sometimes mixed with milk, so that they can meet their nutritional and hydration requirements. It is also used as part of large ceremonies, something I was very humbled to witness. I was present when the community was circumcising boys as they transitioned into warriorhood, and cattle blood was part of the experience. A young man—a warrior—aimed a bow and arrow at the neck of a cow that was being held by another warrior. After a couple of shots, they hit their target and a thin rivulet of blood streamed into a small gourd. Once the container was mostly full, they held the small wound closed to stop the flow. As I walked around the temporary village that was set up for the ceremony, I saw a large pot brimming with cattle blood, clotting as it was being stirred. The event was otherwise full of energy—the circumciser going from hut to hut, warriors demonstrating intense emotions to encourage the boys to show no fear, elders dispensing words of wisdom, and ceremonial blowing of a special horn. I was allowed to stand outside the action and watch from a short distance, never within it—because I was a woman.

I bring this story up because it illustrates two features: first, that blood is an extremely good source of heme iron for Ariaal men. And second, that many things in Ariaal society are gendered. Men are not present when women are circumcised, either.[1] And certain foods are meant for men or women, too. That heme-rich blood I saw while walking through the special circumcision village was meant for men. Dairy, by contrast, is the food eaten by women and children. That's not to say that men do not consume milk, but it's not the domain of men, especially since women do most of the milking. Dairy contains substantial amounts of calcium—great for helping with bone loss, but a hindrance for iron absorption. Combine that with a relative lack of dietary diversity, and you have a recipe for iron deficiency among women and children. And indeed, metrics indicate that women (28% of postpartum women, my data) and children (31.2% of school-aged children, Shell-Duncan and McDade 2005) suffer from anemia and/or iron deficiency.

Gendered foods are not just limited to settled pastoralists who have been marginalized into sparse environments. Gender-appropriate foods are a feature of many cultures. In the United States there are foods that are believed to be eaten more by one gender or the other, at least by advertisers: yogurt commercials are geared toward women. So are "healthy" foods like quinoa and kale. Meat is believed to be the domain of men (Mycek 2018). While of course men will consume fermented dairy products and women, beef, society "assigns" gender to the food, making it seem more appropriate or acceptable to one gender or another. This belief is widespread among European nations and countries that they colonized. This also means that meat consumption—and its heme iron—is gatekept for women in some parts of the world.

In fact, this hegemonic belief of meat as masculine is so ingrained it has contributed to multiple theories about the evolution of humankind. It is widely accepted that increased consumption of animal products, like meat, was a

[1] For a discussion of the genital cutting of women in this area, I recommend reading Bettina Shell-Duncan et al.'s chapter in the volume *As Pastoralists Settle* (Fratkin and Roth 2006).

part of human evolution. Researchers used this idea to develop theories about how human characteristics evolved, such as bipedal walking or larger brain size. In these early theories, getting meat was a man's job—a vivid example of how meat was coded as masculine in the minds of early paleoanthropologists. In the now-defunct Man the Hunter hypothesis, male hominid hunting was believed to explain the evolution of bipedalism, larger brain size, and tool use—all wrapped up into one adaptive package. Men and their meat were the leading force in humans becoming human.

Sally Slocum, in her article "Woman the Gatherer: Male Bias in Anthropology," argues against the Man the Hunter hypothesis, noting that theories like these ignore women and their activities, and center the driving force of evolution as masculine: aggressive and competitive. She notes: "our questions are shaped by the particulars of our historical situation, and by unconscious cultural assumptions" (Slocum 1975 p. 307). Even after Man the Hunter had been debunked, belief in the importance of meat for human evolution has persisted, with many anthropologists arguing that nutritionally dense meat is a requirement for the maintenance of large, energetically expensive human brains (Stanford and Bunn 2001). No wonder that belief in the eating of meat as a "natural" human (male) behavior has persisted, given that scientists have endorsed it. Meat does not have to be a man's food, but we—the public and the scientific establishment—believe that it is.

It is difficult to get a true sense of how meat is gendered outside of places like the United States, because White, Western beliefs about meat are so widespread, distributed via colonialism, media, and globalization. Women in patriarchal societies often take the least nutritious food in the household, meaning they often do not get much heme-rich meat (Johnson et al. 2018). This is shocking given what we know about women's reproductive iron needs. The fact that women are encouraged to eat less meat impacts their well-being, especially in places where non-heme sources like vegetables are not widely available. This is a prime example of how cultural beliefs about food can impact how iron is embodied, and indeed iron deficiency in women reflects not just a nutritional deficiency but is also a statement about their place in society.

In a way, iron supplements are also gendered. Iron is a major feature of prenatal vitamins and is the target of public health campaigns to improve the iron status of women. Supplementation takes many forms: taking an iron pill, using iron cooking implements, and even genetically modifying foods. Global inequalities in iron availability will be discussed more in Chapter 7, and I'll discuss large-scale iron supplementation strategies there. For now, I'll talk just a bit more about the process of taking an iron pill. They are available as both ferrous (Fe^{2+}) and ferric (Fe^{3+}) salts, with the ferrous salts tending to be absorbed better by the body. However, absorption for iron supplements overall is quite low. This means that iron supplement doses are high, which can potentially cause problems. The recommended dose of elemental iron for pregnant women is 27mg and for non-pregnant reproductive aged women is 18mg, with the assumption that about 1 mg will be absorbed each day (US National Institutes of Health Office of Dietary Supplements 2022).

It seems like a no brainer that if you are iron deficient: you should take as much iron as you can to resolve the iron deficiency. After all, isn't hepcidin expected to be low when you are deficient to promote absorption? In reality, the relationship between supplemental iron dosing and iron absorption is more complicated. In experiments where iron deficient participants are dosed with high doses of iron they absorb less iron than expected and experience unpleasant effects (Beard 2000). Side effects of iron supplementation include nausea, constipation, and gastrointestinal upset. While seemingly not serious, these effects are enough to cause many women to stop taking them. And if people will not take iron supplements because the treatment is worse than the disease, it is worth taking a hard look at them. And since I like my hard looks to be anthropological, I like to look at these "side effects" from an evolutionary perspective.

Evolutionarily, nausea and vomiting are protective responses that allows the body to get rid of something that is making it sick. Nausea and vomiting are typical features of pregnancy even without iron supplementation. It is not uncommon for women to stop taking their recommended iron supplementation due to stomach upset. In Chapter 3 we discussed nausea and vomiting as protective mechanisms against harmful effects of iron on the embryo. It turns out that these protective mechanisms might protect non-pregnant, reproductive-aged women, too. A study of dosing frequency and hepcidin levels demonstrates that taking high doses of iron caused a rise in hepcidin levels that persisted for 24 hours, *even when the body is iron deficient*, making the body inefficient at absorbing subsequent doses for the next day (Stoffel et al. 2020). This meant that those on an every-other-day iron dosing schedule had greater improvement in iron status over the study period, as well as less nausea and vomiting. It seems likely that hepcidin protects the body, tolerating iron deficiency to avoid iron overload. And rather than being seen as distracting "side effects," nausea and vomiting should be viewed as protective effects, too.

Unfortunately, this has not made its way into clinical guidelines, with most gastroenterological, gynecological, and hematological societies in countries like the United States and UK recommending daily dosing. The World Health Organization also recommends daily iron dosing of 30–60 mg of elemental iron (World Health Organization 2016). As long as nausea and vomiting are seen as a nuisance rather than a legitimate evolved response to an excessive amount of iron, suboptimal dosing and subsequent rejection of supplementation by women will be the result.

Eating, but not food

There is a special case of consumption that is intimately linked with iron status in women: geophagy. Geophagy is the eating of certain clays, chalks, and dirt, and it is widespread in certain areas of the world. Actually, the entire suite of non-food-eating behaviors, known as pica, is quite common, and encompasses such behaviors as eating ice, starch, and other non-food objects. There are two aspects of geophagy that make it pertinent to this book: (1) geophagy is associated with anemia and iron deficiency, but no one knows exactly why, and (2) pregnant women (who are also more likely to be iron deficient) are particularly susceptible to geophagy.

You may know someone who eats non-food substances regularly but doesn't talk about it. The stigma of geophagy and other pica behaviors is widespread. This leads many to hide the behavior from others (Young 2011). In the United States, pica is considered a mental disorder and is even listed as an eating disorder in the Diagnostic and Statistical Manual (DSM-5), the handbook psychologists use to diagnose someone with a mental illness. In some areas of the world, it is considered a "culture-bound syndrome," meaning that psychologists don't call it a mental disorder in those regions despite otherwise thinking that pica is a disorder elsewhere. A woman eating clay in Rwanda, for example, would be thought to have a "culture-bound syndrome," while a woman in the United States who was eating clay would be considered to have an eating disorder. Interestingly, clay consumption is normalized in certain places in the United States—clays are present in White woman "wellness" spaces in the United States, meant to be used externally or even internally to "detox" the body. Clay consumption is even commercialized as gastrointestinal medicine in some places, and clay consumption has been documented in wild animals, too. This is all to say, eating earth is more widespread than you might think, enough so that evolutionary scientists have spent considerable effort trying to figure out why it happens.

This is a goal of evolutionary medicine—to explain something via evolution that might otherwise be explained as pathological. The psychology community—except in cases where they consider pica to be a "culture bound syndrome"—clearly considers pica to be an abnormal behavior that is best explained as a mental disorder. Since pica is also associated with iron deficiency, it also could be considered a symptom of micronutrient deficiency. However, because geophagy is such a widespread phenomenon and not just a small culture-bound phenomenon, it is a prime candidate for evolutionary exploration. And because geophagy is gendered consumption that is linked to iron deficiency, it is a frontrunner to be related to the embodiment of iron. This is another area where embodiment and evolutionary medicine goes hand in hand, showing how culture, stigma, and other social features can get "under the skin" of our evolved bodies to impact our biology.

In a series of books and articles, Dr Sera Young laid out potential hypotheses to explain geophagy (Young et al. 2011). She dissected multiple possibilities to arrive at three major evolutionary motives: (1) hunger, (2) an attempt to resolve nutritional deficiencies, and (3) protection of the gastrointestinal tract against toxic substances and harmful pathogens. All three explanations seem plausible and difficult to test. Young takes the available information and does her best to find the best support for one of them. In her book *Craving Earth*, she begins by pointing out that geophagy is not just the indiscriminate eating of dirt. Clays, chalks, and starches are carefully chosen for eating. Geophagy is not done at random by any member of the population, either—it is especially common among pregnant women, followed by children. Finally, geophagy is associated with micronutrient deficiency, notably iron deficiency and anemia but also sometimes other mineral deficiencies. How do these three facts best align with the proposed evolutionary hypotheses?

First, let's talk about who has pica and what they eat. Geophagy is especially common among pregnant women. According to Young, geophagy happens pretty much everywhere, to a degree—while the rate of geophagy might be lower in a place like Denmark (0.02% of pregnant women), it still exists. Over 60% of pregnant women exhibit geophagy in some areas of east Africa. Pica might also not necessarily be geophagy. Young reports that in areas of the United States, women might not eat clay or chalk, but instead eat ice or starch (e.g., raw rice, cornstarch) during pregnancy. It is possible that these items are more culturally acceptable or available to pregnant women in regions that do not have socially acceptable outlets for geophagy. Even the consumption of non-food is bound by social convention.

In areas where clay is eaten, not just any clay will do. The clays are often carefully chosen, and clay itself has specific properties that have attracted the attention of medical practitioners. Clays are full of positively and negatively charged ions layered together, which allow them to bind to and absorb many substances, especially when they are dissolved in water (Young et al. 2010). This is why clays are used to contain and absorb waste and chemical spills—they are that good at binding. Clays are also used medicinally, in both the past and present—Kaopectate ("The Diarrhea Specialist™") used to be made with kaolin clay before a formulation change in the United States in 2003.

Second, it is clear that geophagy, and pica in general, is associated with anemia and/or iron deficiency. Young summarizes a series of studies that show that anemia or iron deficiency is the most common nutritional deficiency among people with pica, showing up as "always present" in 77% of populations surveyed that report pica (Young 2011). The directionality of the effect is unclear. While it is commonly believed that pica causes anemia, it is also plausible that anemia causes pica instead. Young's work has found that iron deficiency is the most common micronutrient deficiency associated with pica behaviors. Zinc deficiency might also play a role in pica, but this finding is not as consistent. Calcium deficiency, or at least lack of dairy consumption, has been hypothesized to be associated with pica (Wiley and Katz 1998), but Young tends to discount the role of calcium deficiency in geophagy. I will discuss the possible implication of Wiley's findings in my section on microorganisms.

Based on this evidence, Young evaluated the three alternative hypotheses. The first, that pica is a result of hunger, is quickly rejected. She found that while pica substances might help people feel full, those that eat non-food substances usually do have foods available even when they have pica cravings. The second hypothesis, that pica is the result of micronutrient deficiencies, was a bit more difficult to demonstrate. While iron can be found in soils, it is not very bioavailable, and there is little to no evidence that it is meaningfully absorbed into the body. Usually, iron from soil is absorbed by plants, which are much more easily digested by animals (Young 2011).

After analyzing the literature, Young believed that the last hypothesis—the "protective hypothesis"—makes the most sense. This hypothesis states that pica substances shield the gut from the effects of toxic substances and pathogens. She amasses literature that demonstrates that clays are highly protective against the absorption of plant toxins and pathogens in both humans and other animals who consume clay. It's also quite common that animals and humans use these clays after consuming toxic plants or after having diarrhea. The late Dian Fossey, for example, when studying Rwandan mountain gorillas, noted that they ate great quantities of clay during the dry season when their source of food was bamboo, which caused massive diarrhea. Young also notes that the distribution of geophagy across societies is particularly concentrated in tropical areas that are more likely to have endemic gastrointestinal pathogens. Finally, she also discusses the fact that the humans that are most vulnerable to toxins and pathogens, young children and pregnant women carrying vulnerable fetuses, are the most likely humans to also have pica. Therefore, the craving and consumption of clay is mostly likely to be associated with the protective hypothesis; and micronutrient deficiency, particularly iron deficiency, is a byproduct of the reduced absorption that is a function of eating clays.

Food cravings and aversions, in general, are common among pregnant women—not just clay. It is commonly thought that the craving of substances like clay and the aversion to foods like meat or bitter vegetables are protective against foodborne pathogens and toxins (Fessler 2002). This would make pica a sickness behavior. These are behaviors that seem maladaptive on the surface but in fact prevent the body from making an illness worse or spreading it to others. Fatigue and withdrawal from social activities, for example, seem like negative effects of the flu, but they help the body expend less energy, allowing the its immune system to fight the infection, and prevent the spread of the flu to others. Sickness behaviors are notable because they are spurred on by inflammation—signals from the immune system that usually indicate infection. If pica is an inflammation-mediated sickness behavior, that means that iron withholding (Chapter 2) could be playing a role in inducing anemia when the gastrointestinal tract is inflamed. While the protection of the fetus from the toxicity of iron is one explanation for the presence of pica cravings, it is possible that there is another explanation for these behaviors, one that is rooted within inflammation and the bacteria we share a body with. I will expand on this hypothesis in the next section when I discuss the role of the microbiome in the embodiment of iron.

Conclusion

It might seem strange that a social construct like gender can influence your biology, but indeed gender can be embodied—even via the foods that we eat. In many places, even the United States, gender shapes the foods that we eat and the type and amount of iron we consume and absorb. This has consequences for iron status and the illnesses associated with having too much or too little iron. Gender, too, is associated with the non-foods that we consume—mainly during pregnancy, which is a highly gendered state—and this also can have an impact on iron status. It might seem strange to consider food as a form of gendered embodiment—most examples that are discussed in introductory college courses include cultural practices like foot binding or corset wearing—but cultural perceptions of masculine and feminine foods can have very real effects on the state of the body. Given that women are believed to not eat the recommended daily amounts of iron in many countries, the social role of the foods we consume should not be overlooked. Combined with the reproductive activities most often associated with women's gender—like pregnancy—we can see that lower iron is a perfect storm of both embodied social roles and evolved biological mechanisms.

I would be remiss if I let you believe that embodiment only affected human cells. Our bodies are really a hodgepodge of many organisms, most too tiny to see. These organisms, or microbiomes, also play a role in the embodiment of iron. They both take in iron and change their human host's relationship to iron. The fascinating relationship between the human multiorganism and consumed iron is just beginning to be understood.

What they eat: The diet of your microbiome

Open an elementary school level book of science experiments, and you'll probably see an experiment that demonstrates how digestion works. From squeezing a bag full of bread and water to simulate how the stomach churns food to mixing chemicals to show how gas is made, these experiments are a fun way to visualize something that you can't see (but that you can hear, feel, and smell!). Since they are done with simple materials and chemicals, it gives the impression that digestion is just a series of compartments and tunnels—almost like a hamster habitat—each with its own set of chemical reactions that turn food into energy and waste. Simple, and easy to understand.

Human guts aren't just sterile tubes full of chemicals that break down food. They are also full of life. The microorganisms that live in our gastrointestinal tracts help themselves to the products of our digestion, feeding themselves and producing waste materials that shape human metabolism. It might sound disgusting to have small organisms siphoning off food—*your* food, that you ate for *your* body—but the metabolic products they produce can be beneficial to us, so much so that our body has evolved along with our microbiomes and are inextricably linked. We would die without our microbiomes. Our body is not just a pure temple, it is full of residents and visitors. Humans are cooperative organisms, containing a multitude of individuals representing thousands of species. It's not always one big happy family, though. Some of these microorganisms are helpful, some harmful—and some are in between.

This section discusses the bacteria that live in a human body: both "good" and "bad." Even the "bad" bacteria—the kind that cause infection—use similar mechanisms to bacteria that are "good," and the effects of both are vital to the normal operation of the human body. The following section discusses how iron is utilized by the microbiome and affects the human condition, operating as a pathway of embodiment between society, diet, and iron status. I also continue the discussion of iron and geophagy, situating the microbiome as a mediator between pica behaviors and iron deficiency.

The microbiome

The gut microbiome is a collection of microorganisms—bacteria, fungi, and viruses—that have evolved to live in the gastrointestinal tract of animals. Bacteria are organized using taxonomy, which describes the hierarchical divisions used to describe life on Earth. Just as humans are categorized into the Kingdom Animalia, Phylum Chordata, Class Mammalia, Order Primates, Family Hominidae, Genus *Homo* and Species *sapiens*, so too are bacterial species categorized into different hierarchical levels. It is common to refer to different levels of taxonomic organization to discuss the microbiome. For example, a frequently used term to describe the types of bacteria in a microbiome sample is Operational Taxonomic Unit, or OTU. So the discussion of microbiomes can be confusing if you are not familiar with how they are organized. Before I discuss the effects of the microbiome, I will first discuss the major points of microbiome composition. At certain points, it may be necessary to introduce various levels of taxonomic organization. When this happens, I will make certain to clarify the relationship to higher-order levels.

There are six main phyla of human gut bacteria, containing thousands of species: Firmicutes, Bacteroidetes, Actinobacteria, Proteobacteria, Fusobacterium, and Verrucomicrobiota. While there is no "core" microbiome—in terms of composition of species—that all humans share, there are consistencies between them. Species from Firmicutes and Bacteroidetes tend to dominate healthy adult microbiomes. Actinobacteria, a phylum consisting of species that are often found in soils, are a small but mighty percentage of the human microbiome. *Bifidobacterium*, for example, is a genus within Actinobacteria, and it is frequently used as a human probiotic given its anti-inflammatory and pro-gut barrier functions. Proteobacteria is a phylum that consists of species that instigate inflammatory reactions in the gut. Some Proteobacteria are human pathogens, while others are friendly residents. Subclinical immune activation from "normal" Proteobacteria helps the body prepare the immune system against infections and other perturbations of the gastrointestinal environment. Fusobacteria also activate inflammatory responses, serving a similar role to Proteobacteria (but as a smaller percentage of the gut microbiome). Finally, there is only one species in human microbiomes from Verrucomicrobiota, *Akkermansia muciniphilia*. This species degrades gut mucin, helping maintain gut barrier function and reducing inflammation.

Gut microbiomes vary a lot within and between populations. The biggest driver of microbiome variation is diet, and some broad generalizations can be made about diet and the microbiome. Multiple types of dietary fiber are crucial in maintaining a healthy gut microbiome, whereas saturated fats, sugars, highly processed foods, and salt do not produce beneficial metabolites and promote dysbiosis, which is an imbalance of the microbiome. There are also geographic differences in gut microbiomes that are partially, but not totally, based in dietary differences between groups. Gut microbiomes may also be affected by the microbiomes of those around them (including pets!) as well as the microbiomes present in their environments. This means that microbiomes are social, reflecting social relationships and the nature of built human environments. Therefore, the microbiome is a fascinating pathway of embodiment from the outside world to the inside gut.

While initial studies of the microbiome focused on its composition, more recent work is starting to focus on its function. That is, researchers are starting to learn what the microbiome *does*, not merely what it *is*. To do this, researchers are beginning to quantify genes and their function within a microbiome (metagenomics), measure the activity of those genes (transcriptomics), and assess the metabolic products of the microbiome (metabolomics). Since taxonomic composition, function, and metabolism have some overlap, I will be discussing both composition and function throughout the next section. Please note that composition and function are not necessarily interchangeable, and that the study of microbiome iron function is in its infancy. However, there are broad discussions about the function of the gut microbiome in the presence of iron that can serve as the backdrop to our growing understanding of microbial iron use.

One of the main functions of a commensal microbiome is to produce metabolites that help the gut. These metabolites are short chain fatty acids, or SCFAs, and they are vital for the maintenance of the intestinal barrier. The microbiomes of individuals with various chronic disease states, including cardiovascular disease, metabolic disease, autoimmune disease, and cancer, often have lower production of SCFAs and greater amounts of inflammation. Therefore, alterations of the gut microbiome that reduce the amounts of SCFAs are hypothesized to be the cause of many chronic illnesses. Since there are frequently social disparities in rates of chronic illnesses, the microbiome's ability to create SCFAs could be a useful pathway of embodiment between the social world and the state of the body.

Future work within the microbiome involves clarifying the ecological relationships between bacteria, as well as the effects on bacterial function of non-bacteria gut inhabitants (like fungi and viruses that infect bacteria, which are called bacteriophages). As the research community gradually incorporates new transcriptomic, metabolomic, and metagenomic techniques into their work, our understanding of human microbiomes will exponentially increase. And of course, those of us who straddle the social-biological sciences divide will be able to incorporate these new findings into a more integrated picture of the microbial pathways of embodiment.

Gut inflammation

Food poisoning is a colloquial term for a gastrointestinal infection. It is a common experience: you eat food that seems a little "off," and the next thing you know, your abdomen is sending alarm signals. Evolutionary medicine has hypothesized that the diarrhea that results from these infections is protective—it helps flush out the pathogen attempting to restore the gut to normalcy. It has its downsides, of course: the dehydration that comes with diarrhea can be deadly, especially for infants and young children. It disturbs the normal inhabitants of the gastrointestinal tract, too. Any disruption of the normal makeup of the gut, particularly when it causes it to function in a non-optimal way, is called dysbiosis.

Gastrointestinal infections can be caused by viruses, fungi, bacteria, or multicellular pathogens. These microorganisms are all part of the normal gut ecosystem, but infection can come about from either the introduction of a new, pathogenic microorganism or an overgrowth of an organism that normally exists in the gut. Many of the bacteria that cause gastrointestinal illness are in a category of bacteria called Enterobacteriaceae (phylum Proteobacteria), which tend to provoke inflammatory responses within the gut for their own benefit. The taxonomic family Enterobacteriaceae contains many genera and species such as *Salmonella*, *Escherichia coli*, and *Klebsiella*. The complete taxonomy of Enterobacteriaceae is shown in Figure 5.2. This family contains more than pathogens: "Normal" gut bacteria can be Enterobacteriaceae, too, and they can also instigate inflammation. This means that some degree of inflammation is a typical part of gut physiology. The inflammation caused by commensal Enterobacteriaceae is a Goldilocks scenario: your gut needs inflammation that is not too hot and not too cold but is instead just right. Think of it as a way to keep your gut on its toes (well, if guts had toes!).

Taxonomy of pro-inflammatory bacteria	
Kingdom:	Bacteria
Phylum:	Proteobacteria
Class:	Gammaproteobacteria
Order:	Enterobacterales
Family:	Enterobacteriaceae
Example Genera:	
	Escherichia
	Klebsiella
	Enterobacter
	Proteus
	Serratia

Figure 5.2. Taxonomic classification of many inflammatory human gut bacteria.

Infection, or any overgrowth of inflammatory bacteria, leads to dysbiosis, meaning that the typical ecology of the gastrointestinal tract is altered, usually toward more inflammatory states. Antibiotic use to clear infections can sometimes make the problem worse: it can kill off beneficial bacteria as well as those causing the infection, leaving ecological "space" for inflammatory bacteria to bloom. There are also many other mechanisms that explain the growth of inflammatory bacteria like Enterobacteriaceae, including nutritional changes, changes in the oxygen state of the gastrointestinal tract, the acquisition of metals like iron, and interactions between bacteria, such as the production of antimicrobials and horizontal gene transfer. Pathogens often use available dietary resources, including iron, to expand their population at the expense of commensal residents. Inflammation is one mechanism by which they make gastrointestinal life very comfortable for themselves and very uncomfortable for SCFA-producing bacteria.

Enterobacteriaceae have a few ways to promote inflammation during dysbiosis. Inflammation is a response of the immune system to compounds called antigens. Antigens produced by Enterobacteriaceae include endotoxins (such as lipopolysaccharide or LPS) that live in the bacteria and are released when they disintegrate, and enterotoxins (a form of exotoxin produced outside the bacterial cells) that specifically target intestinal cells. Enterotoxins increase the space between intestinal cells (known as intestinal permeability), which releases more fluids into intestinal spaces, causing diarrhea. Both endotoxins and enterotoxins stimulate immune cells within intestines, causing the innate (non-specific) immune system to be activated. Immune signaling cells known as cytokines are produced in response to antigens, which help trigger inflammatory responses. LPS is an especially critical part of the structure of most gram-negative bacteria like Enterobacteriaceae. LPS also happens to induce a strong immune response from animal immune systems. Therefore, inflammatory responses are always a part of the balancing act of the gastrointestinal system. In the case of infection, the inflammation passes a certain threshold for typical gut function, which downregulates the metabolic activities of non-inflammatory bacteria.

Iron-metabolizing bacteria

Most bacteria need iron for their metabolism. Remember, iron was vital in the metabolisms of Earth's early organisms, meaning that most organisms evolved to use iron to sustain themselves (Chapter 2). Inflammatory bacteria happen to be heavy users of iron, and many have specialized mechanisms to extract iron from their environments. These bacteria compete with their human hosts and each other for the dietary iron that humans eat. That means that excess iron can preferentially affect the growth of bacteria that can best acquire and metabolize iron. Over 300 genes relating to iron metabolism, acquisition, and chemical transformation (oxidation and reduction) have been identified in microorganisms (Garber et al. 2020). This means that the ways microorganisms pick iron out of their environment and use it can be quite diverse. I will focus on a few strategies bacteria use to obtain iron and the resulting effects on their hosts.

One way that bacteria metabolize iron is the production of iron-scavenging proteins called siderophores. Siderophores are secreted by bacteria and other cells in response to iron deficiency and bind ferric iron in the gastrointestinal tract, capturing it for use by the bacteria. The ability of bacteria to sense iron in the gastrointestinal environment is mediated by the ferric-uptake regulator protein (Fur) which controls iron-related gene expression, including siderophores. There are hundreds of types of siderophores; the bacteria species that secretes a specific siderophore has a receptor that will take up the iron-bound siderophore. Once in the cell, the iron will be used or stored based on Fur-mediated gene expression.

Human hosts utilize proteins with a high affinity to iron—transferrin and lactoferrin—that siderophores cannot bind to, keeping iron away from bacteria. However, some bacteria have receptors for transferrin and lactoferrin, which allows them to take up the entire protein-iron complex. Other bacteria also lyse (rupture) red blood cells and extract and take up heme, meaning that siderophores are not the only way that bacteria access host iron (Seyoum, Baye, and Humblot 2021). For the most part, these mechanisms have been studied in the context of harmful pathogens. Iron also seems to increase the virulence of pathogens, making them more harmful to their human hosts. So too much iron in the gut environment has the potential to promote infection and increase the harmful effects of dysbiosis.

The effect of supplemented iron on the microbiome has been studied in humans, mostly children living in resource-poor settings who are often being treated for anemia. Anemic children are more likely to have higher amounts of Enterobacteriaceae than non-anemic children (Muleviciene et al. 2018). And often, iron supplementation and iron fortified foods are associated with greater proliferation of Enterobacteriaceae, such as pathogenic strains of *Escherichia coli*. This is especially common in infants and children living in low resource settings (Sousa Gerós et al. 2020; Paganini and Zimmermann 2017). Higher levels of Enterobacteriaceae might even be implicated in a lack of response to supplementation: one study in urban Peru found that anemic children with a higher level of bacteria from the Order Enterobacterales (which contains the Family Enterobacteriaceae) were less likely to respond to iron supplementation (Dorsey 2020). In another study in Côte d'Ivoire, anemic children had higher numbers of Enterobacteriaceae and lower numbers of the genera *Lactobacillus* and *Bifidobacterium* at baseline, and after supplementation their iron status was not improved. Supplemented children also had an increase in Enterobacteriaceae and reduction in *Lactobacillus* and *Bifidobacterium* and an increase in fecal calprotectin (a marker of intestinal inflammation) compared to children who had not been supplemented (Zimmermann et al. 2010).

In addition to inflammation, some studies indicate that excess iron selects for greater virulence, or harm, in bacterial gut populations, indicating that inflammatory bacteria not only bloom, they cause even more damage to the human host (Sousa Gerós et al. 2020). These effects have not been tested in adults. Therefore, it can be difficult to extrapolate these findings to women, even though they often supplement iron. Pregnant women also tend to have microbiomes that appear to be more dysbiotic, even though the pregnancy itself is normal. What this means for women's microbiome function and iron status is yet to be determined.

While these iron-grabbing mechanisms have mostly been studied in inflammatory bacteria, they might also be used by friendly bacteria that need iron for their metabolisms. The genes for siderophores have been found in non-Enterobacteriaceae bacteria. The extent to which this happens is currently not well characterized in human microbiomes. It's possible that iron-metabolizing bacteria help hosts with iron metabolism. Some evidence suggests that humans might absorb siderophores from friendly bacteria to replete their own stores (Jiang, Constante, and Santos 2008). On the other hand, commensal bacteria might find that the presence of excess iron stresses their physiologies, much as for humans—the iron potentially changes their metabolism, the types of bacteria in their ecologies, and the harm that these new neighbors cause (Kortman et al. 2014). Overall, the relationship between iron-metabolizing bacteria and their hosts may be more complicated than first assumed. Understanding the dynamics of their crosstalk in health and disease would go a long way to understanding how iron status is embodied by the gut.

One exception

Most bacteria on Earth rely on iron for their metabolism. One exception is a well-known set of human commensals: lactic acid bacteria in the family Lactobacillaceae (phylum Firmicutes) and in the genus *Bifidobacterium* (phylum Actinobacteria). These microorganisms can use iron as part of their metabolism, but they do not need it to sustain life. These bacteria are a major part of infant microbiomes but shrink to a smaller proportion of bacteria in adult microbiomes. While the prevalence of lactic acid bacteria in adult guts is about 1–2% or less (Pasolli et al. 2020), their presence has an outsized effect on the overall gut ecosystem.

Despite their small presence in the adult microbiome, lactic acid bacteria like *Lactobacillus* are thought to have played an outsized role in the coevolution of humans and their microbiome. The Old Friends Hypothesis is an evolutionary medicine hypothesis that suggests that humans coevolved with certain commensal organisms that play a role in modulating immune response. This hypothesis predicts that the loss of these helpful Old Friends in human microbiomes could be responsible for multiple chronic illnesses, especially those that involve immune dysregulation (Rook 2010). These friendly gut organisms fall into three categories: soil-associated organisms that are found on foraged foods, helminth worms that are present at endemic levels in the environment, and *Lactobacillus* that are found in different fermented foods. Interventions based on the Old Friends Hypothesis have focused on restoring populations of *Lactobacillus* and other lactic acid bacteria in the form of probiotics and fermented foods.

Lactic acid old friends tend to aid their hosts, including improving iron nutrition. Several studies have found that probiotics and foods that are fermented with Lactobacillaceae species/strains aid iron absorption in the gastrointestinal tract (Rusu et al. 2020). It turns out that at least one probiotic bacteria, *Lactobacillus fermentum*, secretes a molecule that is incredibly good at reducing ferrous ions to ferric ions, the ion that is more easily absorbed by the gut (Vonderheid et al. 2019). An old friend, indeed! However, it is important to note that Lactobacillaceae species (as well as the Bifidobacteriaceae family) decrease under conditions of iron supplementation (Seyoum, Baye, and Humblot 2021), which means that they don't find high-iron environments a hospitable place to live. Despite their ability to help the host absorb iron, there are fewer of them to do so when supplementation occurs.

Gut bacteria frequently produce substances that alter the microbial ecology as a survival strategy. Lactic acid bacteria are no different, and their effects are usually beneficial to human immunoregulation. First, they produce lactic acid when fermenting food, which lowers the pH of the gut. This creates an environment that allows SCFA-producing bacteria to proliferate. Inflammatory bacteria are less fond of low-pH environments, and their numbers dwindle. Lactic acid bacteria will also produce substances like reuterin, which have a strong anti-microbial effect, particularly against inflammatory bacteria (Ortiz-Rivera et al. 2017). These two strategies might be another way lactic acid bacteria aid in iron absorption—by keeping it out of the hands of the Enterobacteriaceae that use it. Lactic acid bacteria also aid anti-inflammatory gut responses and produce butyrate. This helps reduce the levels of fecal calprotectin, a marker of gut inflammation. Interestingly, fecal calprotectin also chelates iron to keep it away from bacteria, which is another example of iron withholding (Nakashige et al. 2015). This reduction of inflammation is another major reason that lactic acid bacteria like *Lactobacillus* are seen as having potential as a therapeutic probiotic for a wide range of inflammatory gastrointestinal illnesses (Saez-Lara et al. 2015).

The metabolic products of lactic acid bacteria may even be involved in iron homeostasis for their human hosts. New research has shown that *Lactobacillus* can sense iron levels in the gut and produce substances that inhibit iron absorption. Das and colleagues found that the metabolites reuterin and 1,3-diaminopropane (DAP), produced by Lactobacilliceae species, help decrease iron absorption by suppressing the production of HIF-2α, a transcriptional factor that helps regulate intestinal absorption by its effects on iron transporters like ferroportin (Das et al. 2020). The scientists also found that these metabolites induce the expression of ferritin in intestinal tissues when iron levels are high and may be a mechanism that aids in iron withholding. They also found that *Lactobacillus* species expanded in the gut during conditions of iron deficiency, suggesting that these bacteria compete with their hosts for iron. They propose that antibiotics that target these species may help improve recovery from anemia. While the results don't fully line up with the findings that suggest that *Lactobacillus* aids in iron absorption, it did find specific mechanisms that implicate *Lactobacillus* in host iron homeostasis, particularly in mechanisms that prevent iron overload. It is likely that the full effect of lactic acid bacteria on host iron absorption is due to multiple intersecting mechanisms. Further testing in human microbiomes, rather than in mouse microbiomes, may help figure out what exactly lactic acid bacteria are doing with iron.

Microbiome hypothesis for pica

The effects of the microbiome on iron status are only beginning to be characterized, although clearly microbial members have a complex relationship to human dietary iron. Some use the iron for themselves and provoke inflammatory-mediated iron withholding, preventing the host from absorbing it; others help the human host obtain iron. Researchers even suspect that microbiomes alter human food cravings, manipulating the human host into providing beneficial nutrients and substances that might promote a favorable gastrointestinal environment for themselves. Based on this, and our understanding of the way that microbiome-mediated inflammation can be implicated in iron absorption, my colleague Achsah Dorsey and I suspect that the microbiome is implicated in geophagy. Including the microbiome as part of the explanatory framework of geophagy might also help explain why anemia is so tightly linked to eating clay.

We propose that geophagy protects against microbiome-mediated gastrointestinal inflammation. This proposal is a slight modification of the protective hypothesis supported by Young, which proposes that geophagy is a protective behavior to calm inflammation due to infection or prevent absorption of toxins. We agree with this, but also expand the hypothesis to include sub-clinical inflammatory states like microbial dysbiosis. Our hypothesis incorporates gut dysfunction that is not due to infection, such as environmental enteropathy—a chronic dysregulation of the microbiome due to the influx of inflammatory bacteria from areas with poor sanitation. It also explains the heightening of geophagy behavior during pregnancy: It appears, at least in U.S. populations, that it is normal for Enterobacteriaceae gut populations to expand during pregnancy. The bloom of Enterobacteriaceae is likely a mechanism for the microbiome to be able to extract more energy to support their pregnant human host, at the cost of some inflammation.

We also propose that geophagy is a sickness behavior. Sickness behaviors are a coordinated set of behaviors that emerge from inflammatory signals, including reduced social behaviors, increased sleep, decreased appetite and mobility, decreased interest in sex, depression, and more, to protect the host from further damage from an infection. It is believed that the interplay of inflammatory cytokines and hormones affect the central nervous system to shift the body's metabolic effort away from seeking social contact, food, and reproduction and toward fighting the infection and preventing the spread of the illness. Sickness behaviors underscore the importance of the field of psychoneuroimmunology—which proposes that there is a relationship between the immune system and the brain. For example, studies have found that low-level chronic inflammation may cause behavioral changes that are unwanted in a normal social setting, potentially resulting in culture-bound mental illnesses such as major depression.

Geophagy is also not necessarily a desired behavior—it is often spoken of as a compulsion, is considered a mental illness in some places, and is heavily stigmatized—but may play a role in protecting the gastrointestinal tract from the effects of inflammation. We believe that this is a fruitful hypothesis for testing, and that both children and pregnant women would benefit from the understanding that geophagy is a sickness behavior. It would also expand the suite of sickness behaviors, incorporating behaviors that may be effective for gastrointestinal infections. Hypotheses that explicitly test the microbiome and the composition and function of bacteria—using both metagenomic and metabolomic analysis—would go a long way to unraveling the complex interactions between geophagic clays and the microbiome. We propose a place to start: understanding how geophagy interacts with dysbiosis and lactic acid bacteria and the resultant effect on iron status.

What effect does geophagy have on inflammatory bacteria? Much of the answer to this comes from the literature on livestock, since clay is frequently given to help gastrointestinal upsets. Research in piglets demonstrates that supplementing with montmorillonite clay reduces Proteobacteria and Bacteroidetes and increases Firmicutes (which contains the family Lactobacillaceae), as well as improves gut epithelial performance (Liu et al. 2020). In mice, feeding a high-fat diet induced dysbiosis; the subsequent administration of bentonite clay reduced the bacterial species responsible for the dysbiosis (Lee et al. 2018). Another study in mice found comparable results, with the outcome that mice fed montmorillonite had less inflammation and less evidence of metabolic disorder (Xu et al. 2017). None of these types of studies, which are largely experimental, have been done in humans—likely because people are not "supposed" to eat clay.

As for why iron deficiency accompanies geophagy, dysbiosis can provide the explanation. The inflammation induced by dysbiosis activates iron withholding mechanisms like those of hepcidin and fecal calprotectin, keeping iron from entering the body. The inflammation-induced iron withholding, not the geophagy, would thus be responsible for lowered iron levels. Both iron deficiency and geophagy are triggered by inflammation, but one does not cause the other. Instead, they are a one-two punch of evolved protection in the face of gastrointestinal inflammation, attempting to alter the composition of the microbiome to benefit the human host.

We propose that lactic acid bacteria like Lactobacillaceae have a role to play in geophagy, too. Researchers are becoming aware of the long evolutionary relationships that microbes might have with clays, since clays themselves have such unique, useful chemical properties that can potentially be involved in the metabolic processes of Life. Lactobacillaceae, particularly *Lactobacillus* species that are associated with mammalian guts, appear to interact positively with various clays. *Lactobacillus casei*, for example, appears to live very well on montmorillonite clay, a geophagic clay (Li et al. 2014). Supplementation with montmorillonite appears to increase the numbers of lactic acid bacteria in piglet microbiomes (Liu et al. 2020) and promote the growth of *Lactobacillus* found in fermented foods, while also reducing the overproduction of lactic acid by these bacteria (Li et al. 2014). While not all clays seem to react equally to lactic acid bacteria, more research on this topic can help resolve the sometimes-contradictory results that have been found.

The growth of lactic acid bacteria in the context of geophagy could be another explanation for the resultant iron deficiency that is associated with geophagy. As lactic acid bacteria populations increase, dysbiosis and the resultant inflammation decreases. However, the iron absorption-inhibiting metabolites discovered by Das and colleagues might decrease the amount of iron that is absorbed by the host, a form of iron withholding that, in theory, protects the host from too much iron. However, when the host already has low iron, such as during pregnancy, this can result in too little iron absorption. Further work is needed to confirm this hypothesis; a start might be to test the effects of geophagy and lactic acid bacteria probiotics on iron deficient and iron sufficient people and see what the resultant change in iron status and microbiome composition might be.

I think this hypothesis can explain the paper by Wiley and Katz (1998) showing an inverse relationship between dairy consumption and geophagy. Geophagy probably isn't an attempt to supplement calcium at all; instead, an entirely different mechanism is driving their findings. Although dairy contains high levels of calcium, which theoretically reduces iron intake, dairying societies frequently ferment their milk into products like curds, kefir, yogurt, or cheese. These societies don't need the Lactobacillaceae-boosting powers of clay: they have a substantial input of these bacteria from the fermented foods in their diet. Those populations that do not have a consistent dose of lactic acid bacteria may consume clay instead. An analysis of the metagenomes of multiple human populations shows that many populations have a small but significant presence of lactic acid bacteria, sourced from either dairy or raw vegetables (Pasolli et al. 2020). The types of foods consumed by each population had a significant effect on the amount and types of lactic acid bacteria available in the gut microbiome. Wiley's data deserves a second look, if only to characterize how much the societies they surveyed ferment food—this may strengthen the claims of the role lactic acid bacteria play in geophagy.

A final note about iron, inflammation, and the gut

Bacteria are not the only cause of iron changes within the gut. Multicellular worm infections (helminth worms) are a big one—the attachment of the worms to the gastrointestinal tract can cause small amounts of bleeding, leading to iron loss. Inflammatory disorders such as inflammatory bowel disease or Crohn's disease can cause lesions that bleed, meaning that anemia is a common finding in these illnesses. Of course, helminth infections and inflammatory bowel diseases also involve changes to the microbiome, but the resulting anemia is more likely due to blood loss than iron withholding mechanisms. Indeed, blood loss from gastrointestinal lesions is a common cause of anemia in patient populations, as mentioned in Chapter 4. I will talk more about the types of social structures that make conditions such as parasite infection and bowel lesions more common in Chapter 7.

Conclusions

Iron absorption is a fascinating, winding pathway of embodiment. The intake of iron gets caught in the nexus of the gut: the central place through which iron is taken in and used by the body's inhabitants. This "Gut Central Station" for food contains a complex ecology of bacteria that all have unique metabolic needs that both help and harm their human hosts. The addition of excessive iron, especially via supplementation, can change microbial composition and thus microbiome function. These changes tend to be proinflammatory, meaning that they alter how the body expends energy and organizes its behavioral priorities to mount an immunological response. They may also induce sickness behaviors like geophagy or nausea and vomiting and trigger iron withholding mechanisms, decreasing the body's ability to absorb iron.

The concept of embodiment assumes that social experience affects the CNS, activating the stress response. In the case of iron absorption, there is little evidence that stress itself impacts the absorption or homeostasis of iron. Instead, there is another path: the embodiment of dietary iron via the experience of eating, which impacts inflammation and ultimately behavior via sickness behaviors. In this hypothesis, iron can impact inflammation, which affects the vagus nerve and the brain, bypassing physiological stress systems and directly impacting the central nervous system. The idea that iron can change the brain via the microbiome and inflammation is an exciting one. This means the iron we eat, via culture-bound (gendered?) foods, can change our brains, embodying the culture and structure that shape how we consume iron.

There is another implication of gastrointestinal inflammation being the cause of sickness behaviors: just as experience can be embodied, so too can the body create experience. Sickness behaviors change the nature of the person's relationship to their social world, meaning that the concept of embodiment is not unidirectional. As Niewöhner and Lock (2018) point out, the biological and the social are entangled, and it is just as important to identify pathways that disperse the body into social systems as it is to study how social systems become embodied. Sickness behaviors may be just that: the way that the body becomes social experience for affected individuals. Geophagy is a compulsive, difficult-to-control act, stigmatizing those who eat clay and changing how they relate to others. We should think of pathways of embodiment as bidirectional or as feedback loops, potentially reinforcing each other.

Bodily states that alter social experience are common (ask anyone with a physical disability), but there is much more work to be done to understand how sickness behaviors change a person's relationship to their social world. Scientists have hypothesized that depression, at least in some cases, may be a sickness behavior—that inflammation-induced social withdrawal and fatigue mimic the signs of depression, and depression-like symptoms are common in those with chronic inflammatory illnesses. Those with depression frequently face barriers: to their relationships, work, and social standing. This may also hold true for a behavior like geophagy, which is also heavily stigmatized. How inflamed people are perceived by the greater social world is a crucial piece of context that is missing from evolutionary descriptions of sickness behaviors. Iron deficiency, with its own effects on well-being, complicates this picture even more.

These iron pathways wind their way through the mosaic, weaving in and out of the bodies of the women depicted there. While their gleaming pathways are beautiful, we now pull back to see their contribution to the larger picture, to a sweeping image of iron entangled with social structure. Our gaze follows a path to a diptych of two women: one White, the other Black, both iron deficient.

6
Blood from a Stone

Anna's father was concerned: she had grown listless, pale, weak, and swollen. Her pulse throbbed if she exerted herself. She refused to eat most foods. He was concerned about her health, certainly; but more importantly, her affliction was keeping away the wealthy men that her father wanted her to marry. He sat down and wrote a letter to a famous doctor, hoping that he would have a cure. As a Renaissance European man of some wealth, he expected that he would receive an answer to Anna's suffering.

Two hundred years later, an enslaved Black woman snuck to a secret place and ate clay that she had hidden away in her skirts. She was also listless and weak, and the palms of her hands were pale. Her stomach was giving her grief, too—*mal d'estomac*, as it was known in Jamaica. She found it difficult to work on the sugar cane plantation where she was enslaved, as her heart throbbed when she did any heavy lifting or walking. The clay she ate was carefully selected on the advice of the other enslaved women. She ate in secret because if her overseer knew she was eating clay, he would call a physician who would prescribe "pukes"—medications that would make her vomit—mercury (which was, of course, toxic), the blistering of her skin, or even the placing of a heavy, locked mask on her face to stop her from consuming the clay. The overseer would also likely punish her, seeing her weakened state as just an excuse to avoid arduous work.

These two scenes occupy the center of our mosaic. Tesserae glitter, adding their light.

Introduction

At this point, I've established that women have an evolved tendency toward lower iron stores and are potentially vulnerable to iron deficiency anemia. The line between health and anemia rests on a knife's edge for women. It relies on everything going well: adequate nutrition, particularly of high iron sources; a functioning gastrointestinal tract with a supportive microbiome, without bleeding from parasites and damage from inflammation; appropriate medical

Thicker Than Water. Elizabeth M. Miller, Oxford University Press. © Oxford University Press 2023.
DOI: 10.1093/oso/9780197665718.003.0006

management of heavy periods and gastrointestinal bleeding—identifying the illness, managing it, and reducing the harms done. In human societies, the ability to access healthy food, sanitary conditions, and medical help are not distributed equally. This makes women's evolved bodies—so efficient at limiting iron intake—vulnerable to low iron if the necessary conditions are not met.

Typically, "mismatch" hypotheses in evolutionary medicine give a relatively shallow explanation of the "new" environment. Our "new" environments—characterized by globalization, stark inequality, structural violence, and capitalistic healthcare systems—are usually discussed in a sentence or two, to keep the focus on evolutionary theory. In this chapter, I would like to dig deeper into these "new" environments, using theories of health taken from medical anthropology, to contextualize iron deficiency/anemia as an embodied affliction that reflects the stratification of society. To do that, I will discuss how iron deficiency anemia has been constructed as an illness, how these constructs have affected women, and how this view has changed over time. I will focus specifically on how identity, particularly constructs of racial identity, can shape the construction and experience of illness, leading to vast disparities in illness experiences.

In this chapter, I will start by focusing on the racial constructs of the United States and its colonizers. This may be a surprise, given that North America represents only 1.4% of the global mortality burden due to iron deficiency (Stoltzfus 2003). However, despite the overall low disease burden, there is tremendous inequality in who is affected. Black women disproportionately face iron deficiency and iron deficiency anemia during their reproductive years compared to White and Hispanic/Latina women. This disparity contributes to unequal maternal mortality and morbidity in U.S. pregnancies. And despite the dire consequences, many women do not even realize they're iron deficient. The way iron deficiency and iron deficiency anemia has been conceptualized by scientific and medical establishments reflects the structural racism that is entrenched in U.S. society.

While the United States has a lower prevalence of anemia than many places around the world, racial disparities in iron deficiency anemia are long term and profound, with poorly understood consequences for Black women's health and well-being. As we will see, iron deficiency and/or anemia has been used to exert social control over women, a control that alternated with indifference to women's fate. Women's iron deficiency anemia can be understood through an examination of the historical and current patriarchal structures in which women's reproduction are embedded. Racial disparities in women's iron deficiency can be traced to the biomedical construction of iron deficiency across time and the structural violence inflicted on Black Americans via enslavement and race science. This history has shaped the way that iron deficiency manifests in present-day racial disparities and intersects with the gendered risk of anemia. Entrenched sexism and racism make iron deficiency a low priority for healthcare, with Black women suffering the most from this neglect. The U.S. biosocial history of iron deficiency continues to have profound effects on women's well-being in the United States.

A brief history of racialization in the United States

Structural racism is a systematic form of racial discrimination that is embedded in the laws, regulations, policies, and organizational behavior within a society. In the United States, structural racism has its roots in the initial European colonization of the Americas. Early forays of European explorers in the late 15th century quickly paved the way for European colonists to first settle and then systematically displace and depopulate Indigenous Americans by violent conflict, disease, and forced migration. One of the colonists' main goals was to use the Americas' natural resources to bolster European economies. As part of that goal, they enslaved people from West Africa to work on large-scale agricultural plantations. Beginning around 1526, the Atlantic slave trade persisted throughout the Southern United States, the Caribbean, and parts of Central and South America until it ended in 1870. Although made illegal in the United States in 1807, ships continued the trade in defiance of the law until 1859. Portugal sailed its last ship of the Atlantic slave trade in 1870 to Brazil, the last country to ban it.

The American Civil War of 1861–1865 ostensibly ended the practice of enslaving Black people via the Emancipation Proclamation. However, after the brief Reconstruction period in which the North attempted to subvert the entrenchment of a partially enslaved status, Southern politicians soon retook state legislatures and escorted in the Jim Crow era. While the Civil War dramatically reduced the wealth of Southerners, Jim Crow policies guaranteed that Black Americans would not be able to build assets the way White Americans could. Jim Crow policies leaned on the 1896 Supreme Court case Plessy vs. Ferguson, which stated that public accommodations such as transportation, schools, and public and private spaces, could be separated by race provided they were equally available, without being in violation of the Fourteenth Amendment guaranteeing equal protection to all. Of course, these separate facilities were not equally available in practice. And further infringement on rights, such as poll taxes and literacy requirements, kept Black Americans from exercising their right to vote, run for public office, or serve on juries. During this time, policies were enacted, such as the practice of "redlining" neighborhoods, that enforced racial segregation and kept Black Americans and other minorities entrenched in certain neighborhoods. The "separate but equal" doctrine was maintained in the United States for over 50 years until the mid-20th century.

Two major federal interventions, the Supreme Court Brown vs. The Board of Education decision in 1954 and the Civil Rights Act in 1965 ended the Jim Crow era on paper. However, the results of centuries of enslavement and decades of unequal treatment persist today in the United States. For example, decades of redlining have perpetuated segregated neighborhoods and schools; even efforts to desegregate schools through busing have backslid in recent years. So too, have efforts to ensure access to polls and the ability to run for office. Law enforcement and the judicial system, healthcare, and other social institutions grapple with unequal treatment of Black and other minority Americans within their systems.

Private and governmental policies were not the only form of structural racism in the United States. The scientific endeavor, particularly physical anthropology, has a long history of scientific racism. Scientific racism is the belief that there is a valid biological basis for racial typologies, and that races can be arranged as a hierarchy (usually with European populations at the top of the hierarchy). Scientific racism is an essentialist belief—that if you can identify a person's race, you can then predict their physical, moral, and intellectual capabilities, as race is an essential property baked into every part of a person. Ultimately, the scientific community through much of U.S. history supported scientific racism, justifying the hostile treatment of Black, Indigenous, and other minority groups that were subjugated. Scientific racism had (and still has) a long reach, influencing all biological sciences including evolutionary, anthropological, and biomedical sciences. While scientific racism has been disavowed as pseudoscience, I will show that these beliefs persist in biomedical thinking, including beliefs about iron deficiency anemia.

The essentialization of sex is a byproduct of scientific racism (Schuller 2018). Scientific sexism, as it were, was also an essentialist belief, permeating much of the U.S. school of evolutionary thought. It was a widespread scientific belief that there were (1) two sexes and that (2) these two sexes were distinct, down to the very function of the cells. While some scientists (e.g., Antoinette Blackwell) spent considerable time explaining that the sexes, while distinct, were equal throughout Nature (Blackwell 1875), others (e.g., Edward Cope) distinctly moved the female sex to an inferior position, framing men as physically active and muscular and women as "disabled" by reproduction and affected by sentimental feelings (Schuller 2018). Scientific sexism and scientific racism were linked by the belief that the evolution of two sexes in humans—both the biological sex binary and the social and cultural roles of men and women—was only achieved by the White race. The ties to scientific racism become even clearer when male treatment of inferior women was considered a racial trait, with more "enlightened" races endeavoring not to subjugate women as their inferiority should allow them to be (Schuller 2018).

While there has not been the same denouncement of scientific sexism as there has been of scientific racism, beliefs about the inferiority of the female sex have faded somewhat. However, essentialist beliefs about sex differences persist to this day within fields as wide ranging as neuroscience, anthropology, evolutionary psychology, and biomedicine. Scientific sexism persists, for example, in the belief that menstrual periods are a pathological cause of iron deficiency anemia rather than an opportunity for iron nutrition to demonstrate homeostasis (Chapter 4). When women do not receive the same standard of healthcare because of this belief, scientific sexism translates directly to gendered disparities in health.

Mismatch, bodies, and society

While scholars of evolutionary medicine look for mismatches between our evolved biology, modern environments, and disease outcomes, medical anthropologists point out that disease itself is socially constructed. Social construction is the production of facts—in this case, biomedical facts—via institutional, sociocultural, or other consensus (Nguyen and Peschard 2003). Essentially, social construction says that a particular reality exists because society and its actors agree that it exists. Medical anthropologists have noted that epidemiological and biomedical illnesses are constructed, in large part by the biomedical establishment, but also via agreement with patients and other stakeholders. These constructed illnesses do not necessarily represent the entire range of affliction affecting a group, nor do these categories necessarily map onto biological phenomenon in a consistent way. Rather, these biomedical categories are a product of the society and the people that created them—and, reflecting the underlying social structure from which they emerged, they can perpetuate inequalities.

Not all illnesses fall neatly into biomedical categories. Illness categories represent how society views the natural world. They are constructed with the full force of existing inequalities behind them. Affliction, by contrast, is more inclusive of the spectrum of human distress. Afflictions are forms of human suffering, regardless of their position in the spectrum of clinical diagnoses. Afflictions, whether biomedically constructed or not, are the embodiment of social inequalities in society. Embodiment, the process of social inequality becoming expressed as a bodily affliction (Chapter 5), can be shown by higher rates of illness and mortality among groups that are farther down societal hierarchies. Nguyen and Peschard (2003) note that:

> affliction must be understood as the embodiment of social hierarchy, a form of violence that for modern bodies is increasingly sublimated into differential disease rates and can be measured in terms of variances in morbidity and mortality between social groups. Ethnographies on the terrain of this neoliberal global health economy suggest that the violence of this inequality will continue to spiral as the exclusion of poorer societies from the global economy worsens their health—an illness poverty trap that, with few exceptions, has been greeted by a culture of indifference that is the hallmark of situations of extreme violence and terror.

Embodiment has been used to explain health disparities in places like the United States. Gravlee (2009) notes that while biomedical practitioners frequently believe that genetic differences underlie racial health disparities, the differences in treatment between racialized groups within a society can cause an individual to be more likely to be in poor health and that the experience of living as a disadvantaged person can be enough to "get under the skin" and lead to worse outcomes for affected groups. Embodiment perspectives rely on an examination of experience, which in social science falls under the umbrella of phenomenology, or the way that things are experienced by our consciousness. While this explanation has been used in the United States to explain racial disparities in health outcomes, its explanatory framework can be translated to other contexts where inequalities persist, particularly the embodiment of poverty. While embodiment has been a useful concept to explain health disparities in medical anthropology, the biological mechanisms of embodiment are more elusive. Diet and stress are two obvious mechanisms of embodiment, with more becoming clear as we learn about epigenetic processes and the microbiome (Chapter 5). This is a compelling mechanism linking structural social forces with human biology, and more examples are needed to demonstrate how the embodiment process creates afflictions that are constructed as illnesses.

Affliction

Iron deficiency as an affliction is not always equally felt by sufferers. Some of the symptoms are subtle. According to the American Society of Hematology (2022), the main symptoms of iron deficiency anemia are "being pale or having yellow 'sallow' skin, unexplained fatigue or lack of energy, shortness of breath or chest pain, especially with activity, unexplained generalized weakness, rapid heartbeat, pounding or 'whooshing' in the ears, headache, especially with activity, craving for ice or clay—'picophagia,' sore or smooth tongue, [and] brittle nails or hair loss." Many, but not all, of these symptoms relate to inadequate oxygen transport to tissues and are sometime only apparent during activity. Apart from picophagia, or pica, many of these symptoms are easy to write off, by both the sufferer and their caregivers. And the symptoms are not always felt equally: active people are more likely to notice symptoms earlier, while those that are less active may not notice symptoms until later.

Pica, by contrast, is a visible manifestation of mineral deficiency. Pica is the craving and consuming of non-food substances, frequently ice, clay, dirt, and chalk (Chapter 5). While not always associated with iron deficiency, there are strong associations between iron deficiency and pica (Chapter 5). Pica is widespread, although frequently stigmatized. Pica is often concealed to avoid disdain, to avoid appearing poor, or to preserve privacy in the face of White male authorities who dismiss pica as an aberrant product of inferior societies (Young 2011). Pica, although a highly visible sign of iron deficiency, is kept secret from authorities—including medical authorities—by sufferers. Why is this affliction, looming large in the lives of those afflicted, not brought to the biomedical establishment? Why is iron deficiency a closely guarded secret among those that are suffering?

The answer to this lies in the biomedical construction of iron deficiency as an illness. The construction of iron deficiency, including pica, as a biomedical diagnosis has a complex social history. One illness construct, chlorosis, is an illness of pale, languid teenagers who are just as sick from unrequited love as they are from a nutritional deficiency. Another, Cachexia Africana, is a disease of enslaved women and children, characterized by dirt-eating and cured with violent methods. The construction of these two illnesses reflects the position of the afflicted women in society. Both conceptualizations are highly patriarchal, and one is racist. The similarities between the afflictions, but the differences in the construction of illness, reflect societal attitudes toward health that still resonate today.

Historical constructions of iron deficiency

Chlorosis, or green sickness

> But, soft! what light through yonder window breaks?
> It is the east, and Juliet is the sun.
> Arise, fair sun, and kill the envious moon,

> Who is already sick and pale with grief,
> That thou her maid art far more fair than she:
> Be not her maid, since she is envious;
> Her vestal livery is but sick and green
> And none but fools do wear it; cast it off.
> (Shakespeare, 1599, *Romeo and Juliet* Act 2 Scene 2 lines 3–10)

The illness chlorosis was first articulated by the physician Johannes Lange in 1554. In a letter to "Anna's father," Lange published the first written description of chlorosis in Western medicine. Lange began by noting that Anna's affliction was keeping wealthy suitors away and that local doctors could not decide on a diagnosis. He then goes on to describe Anna's symptoms:

> You confirm that the character of her face, which during the past year blossomed in rosy cheeks and red lips, has lately turned pale, as if bloodless; her sad heart trembles severely at any bodily movement, and the arteries of her temples pulsate with feeling; she has an attach of dyspnea when dancing or climbing stairs; her stomach turns away from food, above all from meat; her legs—especially near the ankles—swell with oedema at night. (Lange 1554 as translated by King 2004 p. 46)

Fortunately, Lange knows the answer to this medical mystery. Anna is beset with the "disease of virgins." Leaning on Hippocrates's ancient text *On the Diseases of Virgins*, he notes that this illness is caused by obstructed veins, preventing menstrual blood from leaving the body. The menstrual blood fills the abdomen, causing shortness of breath and heart palpitations. Although the symptom profile eventually changed over time—digestive symptoms were added, reflecting an increasing focus on constipation in the 19th century—blocked menstrual blood was always the main cause. In addition to Lange's symptoms—pallor, scanty menstruation, fatigue with exertion, heart palpitations, and abdominal pain—other symptoms were observed such as constipation, dietary disturbance, pica, and poor digestion. Disordered eating, particularly too little food and the excessive consumption of non-food substances, played an increasing role in the etiology of chlorosis over time. A green tinge to the skin was sometimes described—cause unknown—but just as often, paleness was a key symptom. In fact, King (2004 p. 47) notes that the quick acceptance of "the disease of virgins" was likely due to its widespread understanding as a folk diagnosis—Lange states that it is "what the women of Brabant usually call 'white fever,' on account of the pale face, as well as 'love fever', since every lover is pale, and this color is appropriate for a lover." Lange is careful to note, however, that the phenomenon is not actually accompanied by fever.

This "disease of virgins" was variously known as chlorosis or green sickness. References to green sickness abound in Renaissance writings—Juliet, for example, is compared favorably to the moon, who is sick, pale, and green—green sick, in fact, with envy for Romeo's love of Juliet. In writing this, Shakespeare, like others of his time, tied chlorosis intimately to young girls and the promise of their fertility. Chlorosis, as a disease of virginal young girls, was intimately tied to belief in the vanity of young women. Paleness was a standard of beauty in Europe at the time. Jean Varandal, the physician who coined the term chlorosis, introduced the belief that girls made themselves ill to make themselves more attractive: "Convinced of the truth of this maxim that love is synonymous with pallor, young girls and weak unmarried women try by every artificial means to make themselves more pale, in order that they may seem more beautiful" (translated from Latin by Starobinski 1981). Chlorotic girls, then, are hoping for illness, in hopes that they can improve their looks.

So, what was the cure for chlorosis? Lange notes that Hippocrates claims bloodletting as one cure. He does reliably impart this advice, but he also conveys that Hippocrates has a better cure, one that goes straight to the source: "order virgins suffering from this disease to live with men as soon as possible, and have intercourse. If they conceive, they recover" (quoted in King 2004 p. 48). This would allow the veins of the womb to open and allow menstrual blood to flow freely. Marriage, then, was the only socially acceptable way to enact this cure, assuaging Anna's father's fears that she was unmarriageable. King (2004) notes that marriage was the only way to put the chlorotic girl's body under male control, highlighting the male prerogative to impregnate virgins and control their reproductive output. It is no surprise that the treatment for chlorosis was for women to fulfill their societal function as mothers and wives.

While chlorosis as a diagnosis changed through time, it maintained its identity as an illness constructed almost exclusively for adolescent girls. It manifested as an illness that made them meek and submissive—a social ideal—and conferred upon them paleness, a beauty ideal. Its symptoms were romanticized as lovesickness, not because the girls themselves were necessarily lovelorn, but because the illness rendered them into objects of love, waiting to be swept away into matrimony and domesticity. As time passed, lovesickness became paired with neurosis, making afflicted girls' dispositions even more fragile. This did not reflect the reality of chlorosis for many girls. While the typical perception of chlorotic girls was that they were members of the middle to upper classes (Figlio 1978), physicians of the 19th century diagnosed many chlorotic girls from the working and lower classes, particularly industrial workers. At the same time, physicians condemned chlorosis as a manifestation of the refinement and idleness of the well-to-do set, urging girls to stop wearing constricting corsets that kept them from full vigor (Figlio 1978). Even though lower- and working-class girls were increasingly diagnosed at higher rates, they were believed to be aspiring to the behavior of the higher classes and were ignored. Figlio (1978 p. 193) notes:

> With respect to chlorosis and adolescence, my point can be put briefly. Capitalism developed increasingly by calling on youthful female labour. To the extent that the working-class girl was drawn into the labour process, the characteristics of the non-working girl were exaggerated, first, by defining adolescence as a new child-like stage corresponding to the age of intensive labouring in the working class, and then by throwing into ever sharper relief the image of asexual, non-working, delicate femininity.

Chlorosis was a disease of incongruity: when women could not fulfill their prescribed roles, whether it was rich women and marriage or poor women and labor productivity, the specter of chlorosis was near (Wailoo 1999). It was not only that women were afflicted with illness, but the social meaning of the illness was also a defining feature of chlorosis.

In the 1830s, physicians were able to examine the blood of chlorotic girls, learning that chlorosis was anemia. By the 1920s chlorosis was rare and in the 1980s had completely disappeared from medical discourse (Crosby 1987). It is unclear why chlorosis disappeared. It may be that the increase in the use of laboratory science helped diagnose and prevent chlorosis before it became severe; bloodletting fell out of favor with the medical establishment; or, more likely, the illness construct outlived its usefulness as labor laws became more widespread and women's suffrage took hold (Figlio 1978), and Victorian sentimentality for sickly women fell by the wayside (Brumberg 1982). Whatever the reason, by the 1930s it was a rare footnote in medical texts, with women with this affliction being shuffled off to other diagnostic categories. It was merged with "hyperchromic anemia," which was believed to be a syndrome with the same symptoms, but in women that were aged 30–50. The inability to distinguish the older women with hyperchromic anemia and the younger women with chlorosis led to the lumping of diagnoses and the disappearance of chlorosis (Bloomfield 1932). In some cases, anorexia nervosa, a completely different diagnostic construct that involves insufficient food intake, was suspected (Loudon 1980). So, while the affliction persisted, the constructed nature of chlorosis fell by the wayside in favor of new illness constructs.

Cachexia Africana

In the Caribbean and Southern United States, physicians were increasingly mobilized to treat enslaved people of African descent in the 18th and 19th centuries. With the Atlantic slave trade banned by the United States and Britain by 1807, slaveholders increasingly focused on controlling the reproduction of enslaved people, hoping to keep them healthy enough to produce enslaved children (Hogarth 2017). Physicians developed a suite of illnesses that were only applicable to enslaved populations, including a disease they called Cachexia Africana. James Maxwell, a slaveholder and physician in Jamaica, published an article on an ailment that was considered a scourge on Caribbean plantations: *mal d'estomac* or Cachexia Africana (Hogarth 2017). While it was primarily considered an illness characterized by dirt-eating (geophagy), other symptoms included "stomach pains, depression, swelling of the tissues (dropsy), a bloated face, ashen complexion, a failing appetite, vertigo, palpitations of the heart, and shortness of breath" (Maxwell 1835 quoted in Hogarth 2017 p. 81). First originating as an illness in the Caribbean, Cachexia Africana, as a diagnosis, spread to physicians in the Southern United States via medical publications (Hogarth 2017). Physicians used this illness, and others, to paint Black bodies as distinct from White bodies, reinforcing the social and scientific racism of the era.

The eating of dirt and clays was known to physicians in the 19th century who were aware of many examples from past societies in which individuals were known to consume clay, dirt, or chalk (Haller 1972). However, physicians of the 18th and 19th centuries concluded that geophagy was not found in contemporary European peasant societies. Instead, these Southern and Caribbean physicians believed that in the 19th century, geophagy was a phenomenon found among Black populations only. It was believed to be widespread throughout Africa, with Maxwell reporting that "There is not a tribe in Africa [that] did not indulge in eating absorbent clay" (quoted in Haller 1972 p. 239). Such writings established Cachexia Africana as a disease that applied specifically to one race.

Dissent from this view was met with disagreement. The British physician Thomas Dancer wrote in his popular book *Medical Assistant* that Cachexia Africana was no different from chlorosis (Hogarth 2017 p. 95), saying "What analogy there is between Chlorosis and the Pica, I shall not take upon me to ascertain, but diseases which so strongly resemble each other in their symptoms must, it is presumed, have a common cause. The remote causes may be dissimilar, but the proximate one must be the same." But others vehemently disagreed. James Thompson wrote, "When dirt eating occurs in women, it has been compared to the chlorosis or green sickness of Europeans. The comparison is quite erroneous . . . . Green sickness is a complaint, for very obvious reasons, almost unknown amongst young negro women" (quoted in Hogarth 2017 p. 95). He does not elaborate on the "very obvious reasons," but it is likely that the racialized construction of Cachexia Africana is the "obvious" factor he is referring to. In the eyes of physicians, then, Cachexia Africana was a distinctly racialized disease, separate from illnesses like chlorosis that manifest in White people.

Eating dirt, clay, or other non-food items was at first believed to be a social habit of enslaved African people, reflecting a nostalgia for homeland and a wish to die to escape enslavement (Haller 1972). In a sense, it was believed to be a mental disorder due to the trauma of transport during the slave trade. Doctors believed that geophagy would disappear when the Atlantic slave trade ended, yet it persisted past this time among enslaved descendants (Hogarth 2017). At some point, Caribbean physicians blamed the influence of the Obeah religion and its spiritual healers as inciting and perpetuating the habit of geophagy. This was in line with the prevailing belief in Black mental inferiority—Cachexia Africana was one of several illnesses believed to be caused by the hold that Obeah had over enslaved people (Hogarth 2017).

In addition to encouragement by Obeah spiritual leaders, physicians believed that women were also a major force perpetuating Cachexia Africana. Because women and children were more likely to be consuming earth, women were blamed for encouraging the habit among their children. Physicians believed that women engaged in geophagy on purpose, so they could shirk heavy workloads and avoid pregnancy. In fact, historical studies have repeatedly found that White suspicion of enslaved women's behavior stemmed from a belief that they were manipulating their reproduction to resist plantation life (Hogarth 2017). Maxwell believed that breastfeeding for too long was the primary cause of Cachexia Africana, writing, "the erroneous method of nursing Negro children is by far the most frequent cause of dirt eating" (quoted in Hogarth 2017 p. 98). These writings reflect the strongly held beliefs that enslaved women were not fit as mothers because they were feeding children improperly and helping them develop geophagic habits. Further, extended periods of breastfeeding were described by Maxwell as suppressing ovulation and delaying pregnancy, which could be seen by plantation owners as an act of resistance. Therefore, physicians were invested in treating Cachexia Africana as a part of a system that sought to control the reproduction of enslaved Black women.

The treatment for Cachexia Africana was violent. Afflicted people were locked into heavy iron masks to prevent them from eating earth as well as to inflict punishment. They were also aggressively treated with emetics, sometimes daily. In part, these dire treatments reflect the frustration of physicians because their interventions frequently failed to cure Cachexia Africana. Since most believed that dirt eating was an act of resistance, a form of self-harm that pushed back against the plantation system (Hogarth 2017 p. 97), the resulting treatments were frequently cruel, serving as punishment for resistance.

Cachexia Africana, as an illness, was racialized—considered to be an illness in only people of African descent. It was compared to chlorosis, which had similar symptoms (including pica), but this comparison was rejected by the medical establishment of the day. The diagnosis disrupted idealizations about productive Black bodies—the slaveholders' belief that enslaved people could labor under impossible conditions. However, it did confirm physicians' longstanding beliefs that people of African descent were mentally inferior, prone to being led astray by religious leaders and unable to mother properly. Further, Cachexia Africana was believed to be an illness that enslaved women leveraged to resist forced reproduction. The intersectional nature of this constructed illness cannot be overstated. The construction of Cachexia Africana was bound up in racial typologies, beliefs of mental weakness, and fear of Black women's reproductive autonomy. It was used as an attempt to control enslaved people; and notably, the diagnosis did not survive past the 19th century, when Black Americans were emancipated from slavery.

It is difficult to tie Cachexia Africana to iron deficiency anemia—unlike chlorosis, Cachexia Africana disappeared before the widespread use of laboratory blood markers. This means the causative factor for Cachexia Africana is unclear and could potentially stem from a wide variety of causes. Kiple (2002), for example, has suggested that Cachexia Africana was wet or dry beriberi, which is caused by vitamin B1 deficiency. He also discounts the possibility of iron deficiency among enslaved Black people, suggesting their iron cookware buffered them from potential blood loss due to parasite infection; and, in an endnote, suggests there are genetic differences between Black and non-Black populations in terms of iron status (p. 223). According to an archaeologist studying U.S. and Caribbean plantations, the claim that iron cookware use was universal among enslaved people is not supported by evidence (D. Wallman, personal communication). And as we will see later in this chapter, the claim for genetic racial difference is untrue. Further, pica is a common symptom of iron deficiency but not of beriberi, although it is highly likely that more than one nutritional deficiency existed among enslaved people, so that women might be afflicted with both nutritional deficiencies (and more) at the same time, leading to overlapping symptoms. However, the central position of pica in the diagnosis of Cachexia Africana indicates that iron deficiency was implicated in this illness construct for at least some sufferers (see Chapter 5).

Conclusions

Chlorosis and Cachexia Africana are two different, yet similar, illnesses validated by the medical professionals of their time. The exact nature of these afflictions is far from clear from 21st century biomedical standards, but likely involve iron deficiency as part of their etiology. Chlorosis, originally a disease of well-off, young White women, was variously framed as a disease of love, a nervous ailment, and an attempt at vanity. Into the 19th century, both well-off and working women were afflicted, shifting the construction of chlorosis to one where women, no matter their station, could not perform the societal function allotted them (Wailoo 1999). Cachexia Africana had similar features—as the function of enslaved women was to perform labor and reproduce, any affliction that kept them from filling this role was of major concern to plantation physicians. The expression of Cachexia Africana symptoms was viewed as resistance to the designs of the slaveholders and was met with violence. These illnesses vanished once the diagnoses had outlived their social usefulness.

The interpretation of these afflictions—chlorosis and Cachexia Africana—is not uniform among historians. However, one thing is certain: women were disproportionately affected by them, and Black women especially so. The treatment was as bad as the disease: from marriage and bloodletting to masks and punishment. Despite the variation in the reconstruction of chlorosis and Cachexia Africana by historians, one feature is still consistent across both: control of women's fertility. Chlorosis and Cachexia Africana were offensive because they rendered women unable to reproduce, to fulfill the designs of patriarchal societies on their reproductive capacities. This feature of both illnesses is what attracted and kept the attention of the male-dominated medical sector of the time.

The second feature, which was consistent at least among working women and enslaved women, was participation in the labor force. These constructed illnesses rendered sufferers unable to perform labor, which in the case of enslaved women, left them vulnerable to retribution from their overseers. There is less written about anemia and working-class women, but it is likely that non-work in the face of economic need was an undesirable state. Among higher class European women, working was prohibited, so instead the affliction became romanticized. Interestingly, when women's participation in the labor force became more uniform among social classes due to the industrial revolution and war, the illness known as chlorosis melted away. Cachexia Africana, as an illness that was defined to manage enslaved women, disappeared after slavery was ended. In both cases, labor was a key force in changing diagnostic categories.

The disappearance of these illnesses is not as simple as finding better diagnostic methods. As Wailoo (1999) notes, mid-20th century physicians dismissed chlorosis as nothing more than iron deficiency anemia in teenaged girls, forgetting the more complex psychosocial issues that were subsumed under a chlorosis definition. Iron deficiency anemia, now cured with a supplement, became a fact of midcentury womanhood. The new, technological focus on blood as the dominant feature of iron deficiency anemia gave rise to another mid-20th century medical belief: genetic racial differences in iron status.

Into the 20th century: Technological advances and race science

In his book on blood and disease identity, Keith Wailoo notes that the initial work on sickle cell anemia indicated a Mendelian pattern of inheritance, believed to be found only in "Negro blood" (1999 p. 134). Blood became a catchall around which ideas about heredity, kinship, and racial identity were formed. In the early 20th century, hereditarianism was ascendant: the belief that heredity determined human characteristics and human nature dominated the field of genetics. In the early 20th century, these beliefs were tightly linked to race science. Racialized hereditarianism was mobilized to explain a wide variety of human differences and used to justify eugenics practices designed to improve the genetic quality of human populations. Even as scientists learned more about the true nature of human genetic variation worldwide, the belief that racial differences had an underlying genetic basis persisted. In biology and medicine, racialized hereditarianism logic is sometimes still used to explain racial disparities in health outcomes. The more "biological" a racial difference in health outcome appears to be, the more likely it is to be considered genetic. Unsurprisingly anemia, the ultimate marker of inferior blood, has been subjected to scientific racism.

Since the mid-20th century, most scientists and clinicians advocated for separate anemia standards for Black and White individuals. The biological anthropologist Stanley Garn was one of those people. In 1974, Garn and colleagues published an article on racial differences in White and Black hemoglobin measures, pointing out a consistent 1 g/dL difference between groups. Unsure of the cause of this difference, he wrote, "Either there is a real, major and important racial difference in hemoglobin levels, or an equally important but not necessarily genetic difference in nutrition" (Garn, Smith, and Clark 1974). By 1977, he decided that this difference, on the magnitude of 1 standard deviation lower hemoglobin and hematocrit in pregnant Black women compared to White women, was genetic in origin, stating that "it is increasingly appropriate to suggest race-specific norms for hemoglobins and hematocrits" and "The net effect of race-specific norms for hemoglobin and hematocrit norms would be to reduce the proportion of American blacks considered to be 'anemic' or the proportion of black women needlessly given iron supplementation during pregnancy" (Garn et al. 1977). In the guise of helping women, these researchers declined to investigate potential nutritional, socioeconomic, or any other cause of this disparity in hemoglobin and hematocrit. And they recommended that Black women be undertreated for anemia during pregnancy. This recommendation was echoed by biomedical researchers for decades (Johnson-Spear and Yip 1994; Beutler and Waalen 2006). Beutler and Waalen (2006), for example, appear to advocate for race-based definitions of the "lower limit of normal" values of hemoglobin. After excluding 40% of Black women from their reference population for iron deficiency anemia and anemia of chronic inflammation, they conclude that Black women have mean hemoglobin levels and "lower limits of normal" cutoffs of 0.7–0.9 g/dL below those of White women.

Not everyone supported race-based anemia guidelines. In the late 1970s, for example, Dutton (1979) noted Garn's trajectory toward a racial essentialist definition of anemia, stating, "The social implications of a 'separate but equal' definition of disease among blacks are also troubling" and noting that the evidence presented, in its current form, does not do enough to examine the role of economic disparity on anemia vulnerability. Dutton points out that within the samples of children that are used in these studies, Black children are disproportionately likely to be low income and that the shape of the distribution of hemoglobin values differed—there was a left-hand skew among Black children, with more children in the low end of the distribution. Her analysis showed that if Black children had the same average socioeconomic covariates as the White children in the sample, their hemoglobin and hematocrit values would be significantly higher. Dutton also notes that by emphasizing the racial differences in determining cutoffs for anemia over socioeconomic or nutritional differences, fewer Black children would be considered anemic and thus would not be treated. The policy implications of this would shift resources for treating anemia away from Black children toward White children and since Black children already face significant disparities in treatment, changing the cutoffs would exacerbate the status quo. Dutton (1979 p. 954) elaborates further:

> Two points may be noted about the recommendation for separate standards: first, it accords well with the current federal efforts to cut health care costs; and second, it is consistent with the current social attitudes of disfavor toward minority privileges. These parallels are disquieting, for they suggest that concern about methodological rigor may be less if the research conclusions are consistent with prevailing policy directions. In short, the roots of apparently technical consideration may be partly ideological.

For his part, Garn published an article, with colleagues, in 1981 that used an income-matching approach, using it to justify their stance on the "true" racial difference in iron biomarkers that exists outside of socioeconomic status, nutrition, and hemoglobinopathies (Garn, Smith, and Clark 1981). Although it attempted to correct methodological considerations of socioeconomic status, this article neither cited Dutton nor addressed her social concerns.

Taking up the debate, Robert Jackson (1990) noted that while initial studies found a 1 g/dL difference between White and Black individuals in cross-sectional surveys, this difference narrowed to 0.5–0.73 g/dL in subsequent reports that controlled for socioeconomic status, and then to 0.2–0.3 g/dL when dietary intake and iron status was even more carefully controlled for. He highlights the methodological issues with using existing data sets, which over-sampled poor individuals to understand nutritional deficiencies. This means that while the data sets were useful for understanding deficiencies, they were less suitable for examining racial differences within the U.S. population. More importantly, he notes that none of these studies were designed to test genetic differences (via heredity or molecular genetics), so the claims that racial differences are genetic are premature. He also notes that Black populations in the United States are more likely to include carriers for a hemoglobinopathy such as sickle cell anemia or thalassemia, which may cause lower hemoglobin. While this may cause mean hemoglobin levels to be lower for the entire group, it does not justify the lowering of hemoglobin standards. Like Dutton, Jackson argued that the available evidence does not justify a different hemoglobin standard for Black Americans and that setting a lower standard would have the effect of denying treatment to Black people who need it.

The response from biomedical scientists was lukewarm. Nutritional scientists continued to advocate for race-based cutoffs into the 21st century (Beutler and Waalen 2006). Anthropologist Alan Goodman (1996) recognized the racialization of anemia, highlighting Jackson's response to race-based arguments and notes that this was an example of assigning a genetic cause to a racial disparity where no genetic cause has been found to exist. This genetic essentialism was a poor explanation for the differences seen and would have been used to support harmful health policy. The general tide of genetic racial essentialism started to turn in clinical and research communities.

Today, major organizations such as the World Health Organization (WHO) and the American College of Obstetricians and Gynecologists (ACOG) do not advocate for race-based anemia criteria—although in the case of ACOG, race-based anemia cutoffs have only been removed from their published recommendations as of August 2021. The WHO has issued guidance on anemia cutoff thresholds to apply to populations globally—these numbers are based on U.S. populations, with the numbers for women largely extrapolated from White populations (Karakochuk et al. 2019; Sullivan et al. 2008). WHO standards are widespread throughout the literature and are implemented in global public health programs. However, it is unclear how much the old standards persist in clinical practice, and the call for separate standards for women of African origin persists despite the lack of evidence that there is a genetic cause for lower hemoglobin levels that impacts specific racial groups.

Indifference in the 21st century

Sexism and genetic essentialism helped illnesses as visible and as harmful as chlorosis and Cachexia Africana to disappear from the public imagination. Young women's overall lower hemoglobin compared to men and older women allowed physicians to blame their reproduction—menstruation and pregnancy—without investigating other causes for the deficiency (Chapter 4). Scientific studies that started with the assumption that Black people were genetically different allowed different anemia thresholds to persist in the scientific and medical imagination. These stances allowed the medical establishment to regard iron deficiency anemia as an inevitability for women, particularly Black women. The echoes of these beliefs can be seen in today's management of iron deficiency, with or without anemia.

The construction of anemia as genetic is not new; it is merely the current iteration of racial and gender essentialist science. The downplaying of anemia among reproductive-aged women and the separate anemia cutoffs are not a coincidence—this is a form of violence enacted over and over that disproportionately harms certain groups. The fact that iron deficiency anemia is seen as a minor issue—despite the clear harm to anemic women during pregnancy and postpartum—is a part of the problem. In its current form, iron deficiency anemia is an illness constructed by biomedicine, sometimes felt by the sufferer but not confirmed until verified by a biomedically constructed cutoff. Iron deficiency anemia has an existing biomedical explanation and social acceptance of the illness, but the individual experience of the illness is discounted and its symptoms invisible to others. Despite widespread iron deficiency, extraordinarily little literature exists on the experience of iron deficiency. This research is confounded by the fact that the sufferer may not be aware of the illness until she is screened for it. This has long-term consequences for women and their families.

Because the symptoms, which waver between subtle (fatigue) and stigmatized (pica), are not diagnostic, you might expect that women are screened regularly for anemia. Unfortunately, that is not the case. Despite the debate about racialized hemoglobin cutoffs, anemia screening is not generally a part of women's preventative healthcare, at least in the United States. In well-woman visits, clinicians tend to focus on screening the health of reproductive organs, while counseling on matters such as nutritional supplementation is left to the side. That's if the women even show up to a well-woman visit—about 50% of U.S. gynecologists report that they have first contact with patients only after they are pregnant (Morgan et al. 2012). Pre-conception screening is therefore uncommon. It is even less common for minority women, who are less likely to be able to access routine medical care.

This means that more women enter their pregnancies with iron deficiency, with or without anemia. Even when women come in for prenatal care, it is not certain that iron deficiency and/or anemia will be discovered. For example, among women with equal access to Medicaid, minority women were less likely to receive services they initiate or that require specialized care (Gavin et al. 2004). They found that Black and Hispanic women were less likely to receive complete blood cell screening, as well as prescriptions for iron supplements. This affects supplementation behavior: minority women are also less likely to consume iron supplements (Cogswell, Kettel-Khan, and Ramakrishnan 2003).

Anemia during pregnancy is associated with multiple negative outcomes that can cause maternal illness and death. When anemia during pregnancy is severe, it exacerbates the women's birth complications, particularly postpartum hemorrhage, which is the number one cause of maternal mortality. A standard treatment for maternal birth complications is a blood transfusion, which is not easily available in many parts of the world. Women with anemia during pregnancy are more likely to need a blood transfusion after birth (Igbinosa et al. 2020; VanderMeulen et al. 2020). Women who have iron deficiency even without anemia are at risk, although systematic studies are less common. Anemia and/or iron deficiency are both associated with preterm birth and small-for-gestational-age infant size at birth in multiple studies (Rahmati et al. 2020; Badfar et al. 2019), meaning that the burdens of pregnancy anemia are carried by both mothers and their infants.

Iron deficiency post-pregnancy has its own negative outcomes, and in fact the risks of pregnancy anemia, including bleeding, can extend through the first year after birth. Iron deficiency after pregnancy can be subtle. Many of the symptoms are cognitive, affecting the ability to think and even affecting mood. Because pregnancy depletes so much iron from a mother, the period after birth is a vulnerable one for her. In addition to lower iron stores due to pregnancy depletion, she is also coping with a new infant, a change in sleep patterns, breastfeeding, and a new role as parent. While conventional biomedical wisdom suggests that during the postpartum period the risk of iron deficiency in women is at its lowest, due to decreased iron needs and the contraction of red blood cells post-pregnancy, more in-depth work has shown that this is not the case. Most of the research on improved iron status post-birth has been done on higher-income White women, but more recent research has shown that diverse, low-income groups are at risk of iron deficiency post-birth (Bodnar, Cogswell, and McDonald 2005).

Racialization, iron, and depression

In Chapter 5, I discussed how the embodiment of iron could potentially affect the brain. Growing evidence from around the world shows that iron deficiency anemia is associated with worse mental health problems in women, even beyond the compulsive feelings of pica (Chapter 5). In the United States, anemia is significantly associated with depressive symptomology (Lee and Kim 2020). The relationship between depression (especially postpartum depression) and iron status outside of the United States will be discussed more in Chapter 7. In this section, I will discuss the hidden relationships between racialization, inequalities in iron status, and the resultant effect on depressive symptoms in the contemporary United States, demonstrating racial iron inequality and their widespread effects.

My own work, with graduate student Alexis Monkhouse, supports iron deficiency as a contributor to depression in women, particularly among Black women. We used the U.S. NHANES to test this association—which, as a data set, has several limitations for this sort of question. The depression scale within the NHANES, the PHQ-9, is a general screening tool that screens for "likely depression," but it cannot diagnose depression. Nevertheless, it might provide an example of a link between depressive symptoms and iron deficiency. Further, the NHANES itself is extremely limited when it comes to race and ethnicity. The NHANES employs extremely simplistic racial categories—Black, White, and Other (which contains Native American, Asian American, and people identifying as mixed race)—and then a separate "Hispanic" ethnicity category, with those who are labeled "Hispanic" not able to be considered as a member of a racial group. This is not a particularly good definition of race or ethnicity, but it is useful in certain ways. These categories reflect how the United States views the identities of its people. These categories are constructed by the United States, reflects its policies and priorities, and smooths over the very real variation within each category. When the NHANES is used as a research data set, it is essentially a test of U.S. social ideology.

Despite these caveats, we decided to use structural equation modeling to see the relationships between race/ethnicity, iron deficiency, and high scores on the PHQ-9. This is because we could see that the relationships were not just correlated with each other, but instead they influenced each other in a way that suggested directionality between variables. Structural equation modeling makes use of paths in between variables for testing causal hypotheses. We were able to calculate iron deficiency using ferritin and soluble transferrin receptor (Chapter 2) and created a hypothesized model for the relationships between race/ethnicity, iron deficiency, and PHQ-9 score (Figure 6.1). We hypothesized that race/ethnicity, particularly being Black, would be associated with iron deficiency, which would then be associated with higher depressive symptomology. We also saw in initial analyses that body mass index (BMI), a very rough measure of body composition, was separately associated with race/ethnicity and PHQ-9 score, so we moved it to a separate intermediate pathway. We also included some control variables, such as being pregnant and inflammation levels, because of known associations between these variables and iron status.

To test the model, it is run to assess how well the model fits the data and to see what paths between variables are significant. After removing nonsignificant paths, fit is assessed again to make sure it is significantly improved over the initial model. Figure 6.2 represents the final model after significance testing and fit testing. There is a relationship between being Black or Hispanic and greater risk of experiencing iron deficiency, and there is a relationship between iron deficiency and likelihood of higher depression scores (see Table 6.1 for relationship and directionality of all variables). There is also a direct relationship between being Black and risk of higher depression score, meaning

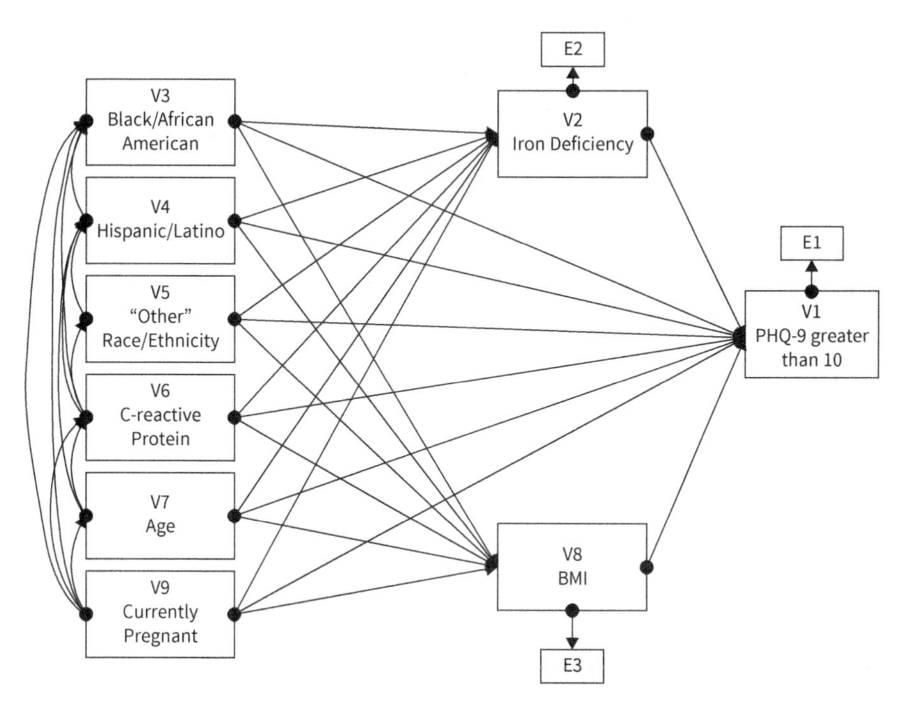

Figure 6.1. Hypothesized structural equation model to test relationship between race/ethnicity, risk of iron deficiency, and risk of high depression scores (PHQ-9).

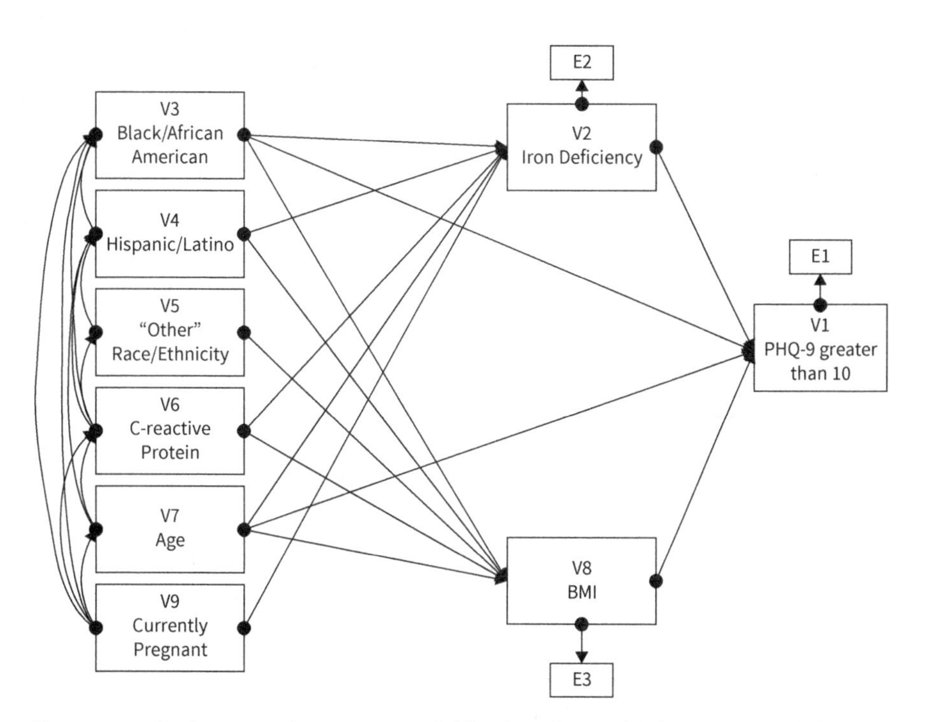

Figure 6.2. Final structural equation model for the relationship between race/ethnicity, risk of iron deficiency, and risk of high depression scores (PHQ-9).

Table 6.1. Estimates, standard errors, and p-values for final model paths. See Figure 6.1 and Figure 6.2 for the structure of the model.

Variable	Predictor	Path	Estimate	Standard Error	P-value
PHQ-9 (V1)	Iron deficiency (V2)	PV1V2	0.04	0.02	0.011
PHQ-9 (V1)	Black/African American (V3)	PV1V3	0.03	0.02	0.0547
PHQ-9 (V1)	Age (V7)	PV1V7	0.02	0.01	0.0026
PHQ-9 (V1)	BMI (V8)	PV1V8	0.03	0.01	<0.0001
Iron deficiency (V2)	Black or African American (V3)	PV2V3	0.07	0.02	<0.0001
Iron deficiency (V2)	Hispanic/Latino (V4)	PV2V4	0.05	0.01	0.0002
Iron deficiency (V2)	C-reactive protein (V6)	PV2V6	−0.03	0.01	<0.0001
Iron deficiency (V2)	Age (V7)	PV2V7	0.01	0.01	0.0173
Iron deficiency (V2)	Currently pregnant (V9)	PV2V9	0.06	0.02	0.0141
BMI (V8)	Black or African American (V3)	PV8V3	0.34	0.03	<0.0001
BMI (V8)	Hispanic/Latino (V4)	PV8V4	0.12	0.03	0.0001
BMI (V8)	"Other" race/ethnicity (V5)	PV8V5	−0.13	0.05	0.0037
BMI (V8)	C-reactive protein (V6)	PV8V6	0.40	0.02	<0.0001
BMI (V8)	Age (V7)	PV8V7	0.09	0.01	<0.0001

that there is a direct relationship between being Black and depression *and* an indirect relationship mediated via iron deficiency. The path between race/ethnicity, BMI, and risk of high depression scores is also interesting, highlighting the multifactorial ways that nutritional status "gets under the skin" and leads to poorer well-being. I would like to note that this is a work in progress; we have not yet incorporated measures of socioeconomic status into the model, which means that the results could change when more social variables are incorporated. Further work may alter the relationships that are depicted here.

Even though this result supports a pathway between being Black, iron deficiency, and potential depression, much more work should be done to evaluate this question. If iron supplementation or an iron-rich diet can help alleviate the symptoms of depression, it is important that this be acknowledged by health scientists and practitioners. This is especially true for Black mothers, who—historically and currently—face the greatest burden of health issues during pregnancy, birth, and postpartum. This work also confirms that, like the historical construction of chlorosis and Cachexia Africana, iron deficiency may also be associated with changes in mental health. This demonstrates another pathway between the embodiment of iron status and altered mental states such as depression. This, in turn, can impact how women relate to their social worlds, heightening the effects and experiences of discrimination.

Unfortunately, the real human costs of iron deficiency and anemia go unacknowledged by science and policy. Even worse, low-income and racial/ethnic minority women disproportionately bear this burden. It is telling that in the scientific literature U.S. racial disparity in mental health is unconnected to disparities in iron deficiency anemia.

Conclusions

Race and the racialization of people has shaped much of iron deficiency in the United States, both historically and currently. You can trace the use of race—what has changed and what has stayed consistent—by following the construction of iron deficiency anemia. From the beginnings of racialized medicine during the U.S. colonial period to the race science that relied on essentialist genetic beliefs about blood during the early 20th century, you can trace how the threads of each weave together to create health disparity in the 21st-century United States. The "racialized medicine" of iron deficiency, like other aspects of race science, essentializes each illness, attributing characteristics that aren't biologically based, like romanticism or willful disobedience, to a racial biological essence. This makes it difficult to separate what is a symptom of affliction and what we have constructed the symptoms to be.

When an affliction like iron deficiency is invisible to the naked eye, it is easy to create essentialist social constructions around it. This makes iron deficiency anemia "mean" different things, and paradoxically can make our true understanding of illness phenomena lag. A slow-growing body of literature links iron deficiency anemia to low energy levels, depression, and mood, demonstrating that its invisible tentacles potentially extend to many areas of a woman's life. Because women are so at risk of iron deficiency, there is potential for it to exacerbate mental health states in depression, particularly in a way that is socially salient. This is just one more way iron deficiency invisibly marks women, heightening the racial disparities on U.S. soil.

Standing back a bit from this corner of the mosaic, we see that these themes—social vulnerability to iron deficiency—are constantly repeated, with a “Where’s Waldo”-like crowd of women depicted in various states of iron deficiency. We look at the border surrounding this mob, which shows a barren agricultural field, the nutritious food dying on the vine. Gazing closer, we notice that the image shows more muted-silver iron pathways linking the Earth’s iron to women’s iron—but that these pathways are disrupted, blocking the flow of iron. What in the world is happening in this scene?

7
Bled Dry

The mosaic is full of smooth lines. Some are silver-gray, winding their way through the image as an iron motherlode. A brilliant burst of lines connects to green, leafy plants, outlining their roots in stark relief. Some lines are small, thin, barely able to be seen between the tesserae. These tendrils barely reach a host of women and children. Lines change from thin to thick and back to thin again as they twist through the image. When we look at the whole mosaic, we can see how our mosaic shows how iron moves from Earth to Life and back again.

Iron flows throughout the Earth. It lies inert, as a mineral, in geological layers of rock. It dissolves into water, becoming part of the oceans and the atmosphere. It is part of the soil, working its way into the structures of plants, which need it to survive. Animals that prey on the plants consume their iron, too—and these animals may be preyed upon in turn, lending their iron to their predators. When the plants and animals die, their iron flows back to the soil. Just as iron flows in and out of the biosphere, so too does it flow throughout the social sphere—flowing adequately for some and inadequately for others. The bio-social life of dietary minerals like iron is a poorly understood phenomenon.

Even though iron is ubiquitous, there is no straightforward path between the Earth's iron and the human body. Multiple biogeochemical cycles describe how iron moves from geological, oceanic, and atmospheric environments to biological entities, and how it returns to the Earth. Pérez-Guzmán and colleagues cleverly describe Earth's organisms as riding "Earth's ferrous wheel" (Pérez-Guzmán, Bogner, and Lower 2010), utilizing chemical, biological, and geological interactions to bring iron into themselves and to return it to the Earth. It's not as easy as eating a piece of iron, or even eating soil containing iron. Our bodies' ability to take in iron evolved in a rich, complex geochemical ecology that was operating eons before the genus *Homo* split from our evolutionary ancestors.

The interactions between minerals containing iron and nature are dictated by the amount of oxygen that is present, the pH of the surrounding environment, and microbes. These factors reduce or oxidize iron ions as they move

Thicker Than Water. Elizabeth M. Miller, Oxford University Press. © Oxford University Press 2023.
DOI: 10.1093/oso/9780197665718.003.0007

within the environment. Humans don't get their iron directly from these reactions—instead, we take in iron that has first been absorbed by plants. Plants require iron to make chlorophyll, and they absorb it from the soil via the tips of their roots. Plants, like humans, are vulnerable to iron toxicity, particularly in soils where there is an abundance of ferrous iron available (Connolly and Guerinot 2002). Humans absorb iron from either non-heme (plant) sources or heme (animal) sources that were metabolized from plant-eating animals. Although these plant and animal sources are widely distributed, the global and local distribution of nutritional iron is not equitable among and within human societies. Therefore, tracing the pathways of iron from the ground to the people that need them is not an easy task.

Since iron makes up about 30% of the world's mass, it is shocking that so many people do not have enough of it within their bodies. The undernutrition of micronutrients like iron is referred to as a "hidden hunger": one that is less visible than the weight loss of caloric deficiency, but no less of an individual and humanitarian crisis. Despite the widespread racial disparity in iron deficiency in the United States, iron deficiency is a minor problem there compared to the rest of the world. Africa and parts of Asia hold 71% percent of the global mortality burden and 65% of the disability-adjusted life years lost due to iron-deficiency anemia. According to Stoltzfus (2003), this represents over 22 million healthy years taken away from people's lifespans due to the disabling and deadly effects of iron deficiency—and these years are taken primarily from women and children. Because of this, much of the public health effort on iron deficiency is aimed at lower- and middle-income countries. However, these efforts have barely made a dent on anemia—estimates suggest that anemia prevalence has only declined by a few percentage points in recent years among pregnant and non-pregnant reproductive aged women (Owais et al. 2021). I will finish by discussing how society has neglected the realities of iron deficiency for women and what can be learned by taking an anthropological approach to this affliction. This chapter will address the global structural inequalities that shape how iron is available or not to broad swaths of the world's women. I propose the term nutrientscape, based on the global flows of culture proposed by Arjun Appadurai, to explain how the biocultural flow of iron is unequally produced, distributed, and consumed across the world.

Global structural inequalities

Globalization perpetuates poverty. European colonialism and neoliberal policies have created a vast worldwide underclass. The World Bank (2022) estimates that almost 900 million people live in extreme poverty, while about 3.4 billion cannot make ends meet. That this underclass is accompanied by a wide range of "public health concerns" should not be a surprise. Large-scale social forces, such as poverty, gender inequality, and racism, determine who becomes ill and who can obtain healthcare. Structural violence is one such way to describe the relationship between social phenomena and ill health. This is a theory of oppression and a form of violence through which social structure or social institutions prevent people from meeting their basic needs, resulting in ill-health (Farmer 2004). Violence is the large-scale disparities in illness severity and mortality that exist due to these structures of oppression—they are violent because they kill, even if no one is throwing a punch or thrusting a knife. There are many ways structural inequalities are hypothesized to create ill-health—through lack of protective social cohesion, the stress of social hierarchies, and material factors. Farmer writes, "An anthropology of structural violence necessarily draws on history and biology, just as it necessarily draws on political economy" (2004 p. 308). An approach that includes history and biology can contextualize how illness and suffering are created and how they become embodied.

The history of global European colonization plays a major role in the story of structural violence in parts of the world. Colonial efforts made many nations and peoples part of the global underclass. In Chapter 6, for example, I talked about how the history of colonialism and enslavement in the United States meant that those in power—including those in the scientific establishment—spent considerable effort to create an underclass of people via racialization. The specifics are unique to the United States, but the widespread nature of colonialism lead to certain similarities across locations. Other nations and peoples across the world were subjected to this treatment, too—the racialization of human groups was an effort, in part, to cast certain groups as "naturally" inferior, thus justifying colonists' poor treatment of them. While racialization is a major social force shaping social position, poverty is a powerful force shaping inequalities.

Poverty is a key feature of critical biocultural theory. This theory is a newish elaboration on the "New Biocultural Synthesis," which introduced the idea that human biology can be understood as a combination of political economy and human adaptationist approaches to the environment and their resultant effect on human biology (Goodman and Leatherman 1998). The political economy of many nations favors inequalities in wealth, sometimes incredibly large inequalities. The inability of those in lower classes to meet their physical and mental needs contributes to alterations of bodily homeostasis and subsequent health inequalities. While this approach is similar to the structural violence approach described by Farmer, it focuses on the effects of social class on human biological processes, offering an intermediate step between social structure and health outcomes. The typical approach of the New Biocultural Synthesis is to describe the political economic structure and history of a nation or place and then describe the biological outcomes, sometimes testing the relationship between the ability to access resources and the resultant biology (such as child growth, health outcomes, or stress response). Studies that use this framework have confirmed the solid links between poverty, human biology, and ill health.

A critique of this approach is that it is reliant on political economic social structures and how they've been shaped by the history of nations to explain biological variation on a local scale. While this does affect human biology, it may not best describe how certain aspects of culture—such as beliefs, meanings, customs, and habits—impact the body; and it does not necessarily provide the most useful framework for understanding global phenomena, which frequently rely on complex interrelationships at different levels of social organization. In particular, it doesn't provide a straightforward way to describe or measure the interconnectedness of places, focusing instead on the uniqueness of each place. When globalization creates circumstances in which human culture flows between and within places, a more nuanced view is needed to understand the local, the global, and everything in between—particularly when it comes to a widespread global phenomenon such as iron deficiency.

The anthropologist Arun Appadurai (1990) created a theory of culture that explains how cultural features moves around the globe, defining certain "-scapes" as being particularly notable: ethnoscapes, financescapes, mediascapes, technoscapes, and ideoscapes. Ethnoscapes comprise the flow of people; financescapes, money; mediascapes, media, particularly images; technoscapes, technology and ideoscapes, or ideas and ideologies. The flow of these -scapes is global, diffusing locally and being changed by local contexts, and then flowing elsewhere. The -scapes themselves are constructs, perceived by actors that have their own unique situated contexts, and help shape identities. Critiques of this idea note that it does not consider the structure of nations, the relationships within and between them, and how these structural forces constrain the flow of these various -scapes to and from certain people and groups. However, the -scape concept is a useful one to help explore how cultural features move throughout the globe and can be helpful when assessing the role of globalization on the topic of interest.

While Appadurai initially created -scapes to discuss the flow of culture, it has also been used to frame biocultural concepts like bodyscapes (Geller 2009). Bodyscapes are dominant ideas about the ideal body that shape how we understand health, gender, and sexuality. They are social constructs that frame the nature of the human body and its identities, and they impact how we understand and relate to human biology. Biocultural concepts like bodyscapes may be a way to link material objects like iron minerals with cultural concepts like nutrition—if we understand nutrition to be dominant ideas about ideal nutrition and the best ways to nourish bodies. If this is true, the flows of ideas about nutrients, the nutrients themselves, and the resultant state of the body would be part of a "nutrientscape" that flows throughout the globe.

Since -scapes, as Appadurai defined them, are sociocultural constructs and not necessarily material objects, the concept of a nutrientscape does not exactly map onto his theory of globalization. However, since the definition, use, and movement of foods around the world is constructed by humans and is constrained by human politics, history, and economy, nutrients are both biological and cultural—essential to the body and also a construct used to understand what the body needs to operate "correctly." Nutrition, and nutrients, also shape how society organizes itself, representing a bi-directional, biocultural concept. And since nutrients—particularly globally distributed nutrients like iron—do flow from place to place, a social theory like -scapes helps conceptualize their movements and define how they are organized in time and space. The movement of iron from nutrient-rich geological "spheres" to nutrient deficient bodies is an important part of conceptualizing the global scope of iron deficiency.

Understanding the nutrientscape of iron can help us understand why dietary iron is not distributed equally and how its distribution is impacted by, and contributes to, poverty. In this chapter, I'll discuss both the structures and the flows that explain the production, distribution, and consumption of dietary iron and how inequalities in iron intake are created. I will certainly not disregard social structure—although Appadurai's theory is post-structuralist—and will use both social structure and -scapes to demonstrate how structures help create poverty while culture and biology—like nutrients—flow within and between them.

I would like to add a little caveat here: since I am not primarily trained in social theory (I would say my social science training is secondary or even tertiary to my training in evolutionary theory and biological anthropology), please take my theorizing with a grain of salt. However, since social theory exists to help guide the understanding of social phenomena, I hope that this approach is useful to help explain how globalization, in addition to local history and political economy, can be used to understand biocultural phenomena at a scale that is commensurate with the global scale of iron deficiency. Big problems require big theory, and often more than one!

Nutrientscape

The location of iron on the planet can better be described as a process than a place. All living things have ways of accessing iron from the Earth—many non-animal residents of the biosphere can access iron from the non-biological spheres of earth—the atmosphere (air), lithosphere (land), or hydrosphere (water). Many animals, by contrast, access iron almost completely from other residents of the biosphere (Earth's organisms). Animals consume plants or other organisms and absorb their iron along with the rest of their nutrients. Ultimately, even if an animal mostly consumes other animals, most dietary iron comes from plants. The process of iron moving from soil to plant to animal to the human body tells us a bit about how we access iron and why some of us humans don't have enough of it.

Nutrientscape of the past

Humans get iron via their subsistence strategies, in the present and especially in our evolutionary past. There are three main "small scale" subsistence strategies that anthropologists like to discuss: hunting and gathering, pastoralism, and subsistence agriculture. These subsistence strategies put humans close to ecological sources of food—the biosphere—and represent a starting point to understand how iron makes its way into the human body.

Evolutionary anthropologists tend to agree that the earliest human subsistence pattern was some form of hunting and gathering. Hunting represents the extraction of animal sources from ecologies, such as hunting large game, trapping small game, fishing, or catching insects. Gathering represents the ability to forage a wide range of plant foods from the environment, and it sometimes involves animals or animal products as well (edible insects, honey). Although these activities might appear simple to the uninitiated, they are anything but—developing the technology and knowledge base to best extract nutrition from the environment is something that takes years of life (and hundreds of thousands of years of evolution) to successfully pull off. The iron available to hunter-gatherers is in the form of both heme (animal) and non-heme (plant) foods. Generally, scientists believe that most calories in most ecological settings of the human past would have come from plant sources, since hunting, particularly hunting for large game, is an inconsistent source of food. Further, human physiology seems to favor increased fat stores compared to other apes, suggesting that humans are buffered (at least calorically) against ecologies that are less productive, food-wise. There is no information about the consumption or buffering capacity of stored iron in humans' evolutionary past—it is assumed to be adequate, but humans' potential for a thrifty metabolism suggests that food sources, including the iron within, were not always consistently available to humans in their diverse past environments.

About 10,000 years ago, humans began manipulating the ecologies around them in a particular way, choosing variants of plants and animals that they could more easily extract food from. Plant and animal domestication emerged in multiple places around the world, creating species that produced more nutrients for humans, but required human intervention to reproduce. Plant and animal domestication introduced new subsistence strategies: pastoralism and agriculture. Pastoralism relies on having open grazing areas for animals but can be flexible in extreme environments that are unsuitable for farming, such as cold (reindeer), arid (camels), and high-altitude environments (llamas, alpacas, yak, sheep, goats, cattle). Farming relies on the domestication of plants, with the chosen variants requiring help to grow and reproduce in the form of planting, watering, and harvesting crops. This requires settling in one place, living near the source of food, and finding ways to prepare and store the food when crops are no longer in season. Of course, mixed farming, which combines animal husbandry with growing crops, is also a possible subsistence strategy in recent human evolution. And sometimes foraging or hunting would accompany farming and herding in certain places. However, despite these new, seemingly successful forms of subsistence, anthropologists have proposed that these newer modes of food acquisition have made humans more likely to become unwell. This includes the possibility that subsistence strategies like farming have introduced widespread iron deficiency to human populations.

Shouldn't iron be easier to access with domestication of plants and animals? After all, didn't it create a more stable food supply? Bioarchaeologists have discussed this problem in the archaeological record. Bioarchaeologists study human skeletal remains, along with the cultural context of their graves and other archaeological evidence. Some graves are linked to a particular subsistence strategy, allowing bioarcheologists to make inferences about the health of the skeletal remains and the societies that the deceased individuals lived in. This means they are well-situated to discuss the relationship between social structure and biological processes that may have affected the skeleton in the past.

Bioarchaeologists have found that there are two skeletal defects that might be associated with anemia, cribra orbitalia and porotic hyperostosis. Cribra orbitalia are lesions that are found on the roof of the orbit, or eye socket. Porotic hyperostosis appears as small perforations or pits on the cranium that make it look porous. Both have similar etiologies, are indicative of nutritional stress, and are surprisingly widespread in skeletal populations. However, the cause of these lesions—iron deficiency or no—has been a matter of debate in the bioarcheological research community.

Anemia is a major cause of these skeletal lesions, based on the dynamics of bone marrow health, red blood cell formation, and the physiology of bone formation under anemic conditions (Walker et al. 2009). In the case of anemia, the bone marrow expands to produce red blood cells, causing the lesions to form. However, anemia is not always due to iron deficiency. While iron deficiency is the leading cause of anemia worldwide, there are other conditions and nutrient deficiencies that may cause it, such as vitamin B12 deficiency, neoplastic (abnormal cell growth) conditions, infectious disease, or injury. Researchers have also noted that the state of the bone marrow in adulthood means that these lesions may not be due to anemia in adults over the age of 30–40, although anemia-related lesions might occur elsewhere in the skeleton such as the vertebrae (Brickley 2018).

The prevalence of these lesions in skeletal populations is high. In a review of skeletal populations in the Western Hemisphere, Walker and colleagues (Walker et al. 2009) found that their prevalence ranged from 20–40% in individuals who died as children and 10–20% in those who died as adults, with prevalence <10% in individuals who died over the age of 60 (without stratifying by presumed sex of the individual). The prevalence of these lesions is even more shocking considering a recent review has shown that they are rare in modern populations—about 0.08% of skeletons in samples measured about 100 years ago show evidence of porotic hyperostosis, the lesion that is more definitively linked to anemia (Cole and Waldron 2019). Walker et al. (2009) suggest low resource availability and inequality in food availability were associated with these lesions.

In general, a survey of skeletal samples showed that health is worse for paleopopulations who lived under domestic agriculture regimes, or in ecologies where nutritional resources tended to be sparse such as open grasslands (Steckel et al. 2002). While the cause of the lesions is unclear, it seems that nutritional deficiencies that are associated with anemia, including iron and B12 deficiency, are reasonable possibilities. This means that micronutrients like iron were not necessarily equally distributed between individuals in the human past—subsistence strategies and social structure played a role in the nutrientscape of past peoples.

Nutrientscape of global foods systems

While anthropologists tend to talk about these different modes of subsistence in idealized terms, even waxing poetic about how our bodies ideally evolved in hunter-gatherer, small-scale settings, it was probably not always the case that early human environments were idyllic nutrientscapes, full of any foods that a human might want. First, studies that are based on living groups of hunter-gatherers or pastoralists do not always mention that these groups are frequently marginalized within their countries (and sometimes, the countries themselves are marginalized on the global stage). The Ariaal people, discussed in Chapters 3 and 4, are settled pastoralists and subsistence farmers. While their pastoralist mode of subsistence is well-suited to the arid conditions of northern Kenya, their status as pastoralists and their location in a rural, remote part of Kenya isolates them from market resources, relief efforts, and healthcare. This is reflected in their health outcomes, including their iron status (Chapter 5). Small-scale subsistence strategies are also vulnerable to environmental degradation and climate change. Even though the Ariaal use a subsistence strategy that places them close to ecological sources of iron (animal sources of food), iron does not flow to them the way that it should.

This is in part because today much of the human food supply is industrialized, relying on technology to produce high yields of food via agriculture. The development of agriculture has created a multitude of food systems that have increased in complexity in the face of technological advances and globalization. Growing human food has become highly technical and centralized, with large growing operations shipping massive amounts of food to other locations for consumption. These operations take place on the best lands for farming, while those using other forms of subsistence, like pastoralism, are pushed to less ecologically productive spaces. While industrialized food production has theoretically been effective at increasing the available calories for the growing number of people on earth, its success at mobilizing nutritious food to every person in the world is mixed.

Food systems, particularly the movement of food from agricultural areas to other places, form the basis for more complex social structures, such as city-states, nations, and global systems. Currently, food systems are dizzyingly complex. Researchers, for example, mapped the supply chains of ingredients for one U.S. fast-food hamburger. They found that the hamburger had more than 50 ingredients that were sourced from every continent except the Arctic (Hueston and McLeod 2012). This does not include the vast number of resources needed to maintain high-tech agricultural systems, from fertilizers to planting and harvesting technology to livestock medication and vaccines. Although agricultural yields can be quite high, including yields of iron-rich sources like meat, these foods are not equally accessed by everyone.

Despite the technological advances in growing large yields of food, many people on the Earth do not have adequate, nutritious food to meet their dietary needs. This is known as food insecurity. This means that food insecure people might sometimes have adequate food, and sometimes not—this is a much different picture of hunger than media images of starving people might suggest. We often think of food insecurity as just not having enough calories, but not having enough of a micronutrient like iron would also be a consequence of food insecurity. This is a "hidden hunger": inadequate micronutrients to fully meet the body's needs. Additionally, many of the symptoms of iron deficiency are hidden, and many of the methods to rectify it are insufficient to meet the needs of the global hungry.

A major issue facing the world's "hidden hungry" is that the types of foods that are affordable and accessible do not tend to be rich in iron. These foods tend to be staple foods: cereals such as corn/maize, rice, millet, wheat-based foods; roots and tubers such as potatoes, sweet potatoes, and yams; and legumes such as soybean products or lentils. These are cheap, easy to cook, and filling. However, they do not have much iron in them—as plant sources, they have less iron and less bioavailable iron for human consumption. They tend to be on the low end of the spectrum for iron content in plants, too—nuts, and leafy greens, for example, have higher iron content than rice or corn. Nuts and leafy greens also tend to be more expensive and less filling, meaning they are not a large part of diets of the poor. Animal sources of iron are even more bioavailable, but also pricier. Further, the combination of which foods are available and where is regionally and culturally bound, meaning that some populations may be more buffered against iron deficiency than others. For those in areas with extreme food shortages—famine—the foods used as relief foods tend to be cereal staples. These foods tend to focus on alleviating hunger but do not always provide the complete nutrition that can alleviate hidden hungers. The iron nutrientscape of famine, food insecurity, and other resource shortages may not flow to the hidden hungry, even as food is transported into stricken areas.

Compounding this problem is the fact that the plants that are grown for food need food too. This food comes from the soil, and the repeated use of soil for agriculture can lead to soil nutrient depletion (Tan, Lal, and Wiebe 2005). Iron is notoriously difficult for plants to access from soils, and this is especially so with iron deficient soil (López-Millán et al. 2013). Plants have multiple strategies for absorbing iron from the soil. Iron uptake by plants, for example, can be modified by adding either phosphorus or substances that help release iron from the soil (Venuti et al. 2019). However, some places, particularly small-scale horticultural farms, may not have the technology to manage soil depletion, meaning that the plants that they grow do not have enough iron and in turn do not provide iron for the humans that consume them. Even in high tech agricultural settings, historical records show that the yield of iron and zinc in staple foods like wheat have decreased over the past 120 years, even with massive increases in plant yields (Murphy, Reeves, and Jones 2008). This suggests that there is a trade-off between increasing the yield of a plant and its subsequent micronutrient content. Certain cereal plants like wheat, rice, and barley have also been shown to absorb less iron when there is more carbon dioxide in the atmosphere, which means that anthropogenic climate change that increases carbon dioxide emissions can also have shocking impacts on the human nutrientscape—specifically reducing the micronutrient content of food even when food yields remain steady (Smith, Golden, and Myers 2017). These changes are unknown to most people, placing an even greater number of people at risk of iron deficiency as the effects of climate change increase.

Given that plants, like humans, also have physiologies that restrict the absorption of iron, there are limits to how much iron there can be in staple foods. Agricultural scientists have begun to address the limits of plant iron absorption using biotechnology—namely, genetic modification. Genetically modified rice, maize, beans, and wheat are leveraged to increase the amount of iron and zinc in them through a host of genetic modifications and crop rotation strategies (Balk et al. 2019). This is known as iron (and zinc) biofortification. Another genetic plant modification is to reduce the phytic acid content of staple plants such as maize, wheat, and sorghum. By reducing the phytic acid, which decreases iron absorption in the human GI tract, scientists hope that more iron can be absorbed from dietary sources. While this technology is exciting, there are still barriers to its use, including cost, safety, and acceptability of genetically modified foods.

Iron biofortification is a new technology. Traditionally, public health officials have turned to fortification and supplementation to fill the gaps left by poor dietary iron in food sources. Iron-fortified foods, supplements, and cooking utensils are all strategies that are leveraged to fortify iron in the diet. Iron fortification is the addition of iron to foods. Sounds simple, but iron fortification is stymied by the same problem as regular dietary iron—it's not very bioavailable, meaning not much is absorbed. Iron fortification of wheat flour is mandated in 81 countries, meaning that in theory, the issues that might arise from iron fortification should have been ironed out long ago (pun very much intended!). In practice, there are many barriers; and a review of 78 national flour fortification programs showed that only 9 could be expected to be effective (Hurrell et al. 2010). Some of the technological issues relating to iron fortification include the addition of other compounds, such as Vitamin C, to make iron more available; the choice of iron mineral; the encapsulation of iron to maintain food palatability and acceptability; the appropriate estimate of fortification level; and the correct biomarkers to accurately measure iron sufficiency (Hurrell 2021). The taste and color changes that iron fortification make to salt, for example, makes it an undesirable target for food fortification.

Beyond this, there are major challenges in making sure iron fortified food gets to the people that need it, including lack of availability of iron fortified foods, weak enforcement of fortification in countries where it is mandated, and inadequate consumption of fortified foods to relieve iron deficiency (Hurrell et al. 2010). Further, because of the highly processed nature of fortification, it pushes people to rely on processed foods rather than traditional foods, potentially increasing the prevalence of chronic illnesses due to high-carbohydrate, high-fat processed foods. While fortification and biofortification are hailed as the solution to global hidden hunger, they would also alter the nutrientscape of traditional diets in ways that are not fully clear.

Supplements are another tool in the public health toolkit. Supplemental iron is a key part of iron nutrition during pregnancy, with the World Health Organization recommending 30–60 mg per day for anemic women (Chapter 5). However, women's use of iron-folate supplementation is strikingly low, falling as low as 1% in some countries and averaging 21% in low- and middle-income countries surveyed (Hodgins and D'Agostino 2014). The ability to access recommended supplementation is dependent on access to healthcare, which is limited in impoverished areas. Public health officials have tried other ways to increase iron supplementation in deficient areas, including the use of iron cookware or iron ingots that can be added to existing cookware to cook with food. However, the use of these items may be intermittent, and their safety and efficacy has not fully been demonstrated (Alves, Saleh, and Alaofè 2019).

Even with the possibility of fortified foods and the genetic modification of plants, it is obvious that the true cause of hidden hunger and insufficient nutrientscapes is poverty. Even if all staple crops were genetically modified to be more nutrient rich, the poor still rely on these foods to be made available to them at a price they can afford (and many can afford nothing). Unsurprisingly, the impoverished are also the hungry. A recent map of the global "hidden hungry" shows that the effects of micronutrient deficiency, particularly iron deficiency, tend to be equatorial, with particularly high levels in sub-Saharan Africa, South Asia, and the Middle East (especially India), and some parts of South America (Muthayya et al. 2013). Muthayya and colleagues' Hidden Hunger Index (a measure of micronutrient deficiency, including iron deficiency) was highly correlated with the Human Development Index ($R = -0.88$), even more so than the Human Development Index and inadequate dietary energy ($R = -0.54$). This means that less economically developed countries carry a greater burden of hidden hunger, highlighting the role of a country's economic development in the well-being of the people within its borders. Poorer countries equal poorly nourished people—people that need better food, not merely more food.

That the "better" food is automatically considered to be bio/fortified is an example of the belief in scientific expertise to solve food insecurity above all other solutions. The term nutritionism refers to the belief that food is primarily a means to deliver nutrients, one that can be scientifically manipulated and optimized to cure hunger. Nutritionism does not elaborate on the other ways humans and their bodies engage with food, including the social contexts of food (Kimura 2013). Nutritionism, as a social construct, is a way that power is exerted over the problematic, malnourished bodies in low- and middle-income countries—using scientific expertise to discipline the poor, making them subject to expert recommendations that know what is best for them. Further, the expert view of the nutritional value of foods has become a clever marketing scheme, meaning that "good" and "bad" foods are marketed to people by their nutritional content, whether or not they are highly processed. Food processing techniques like fortification and biofortification will never provide full solutions to iron deficiency because they take control of the nutrientscape away from the people who need food and put it into the hands of capitalists, hiding it behind patents and proprietary processes. While fortification may help in some ways, it is also yet another way to turn the suffering of the global poor into a profit.

As long as there are extreme inequalities in wealth, the nutrientscape will be similarly unequal. A more complex, inclusive view of the nutrientscape of iron must include not only the highly technical, scientific aspects of nutritional content, but the ways in which nutrients are experienced and consumed by people via food. As I will discuss in a later section, women's access to iron, whether processed or not, is deeply constrained by gendered inequalities that exist across the globe. Women's status as secondhand citizens of capitalism greatly impacts their ability to access the iron they and their families need.

Parasite lost (iron): Nutrientscape of iron loss

Humans are not just inert collectors of iron—we also lose iron back to the biosphere. While normal iron losses via stool are small—remember, iron is difficult for the body to get rid of—there are certain circumstances that cause more regular blood loss. No, not menstrual periods, of course: GI disease and intestinal parasites can cause chronic intestinal blood loss. The flow of iron out of humans is exacerbated in areas where intestinal dysbiosis and infection is part of daily life. Unsurprisingly, these areas are also where much of the world's poor live.

The iron nutrientscape of less developed areas is shaped by the fact that many who live in poverty reside in places with poor sanitation and infection control. This means that they are vulnerable to iron loss via blood loss from parasite infection, GI infection, and lower iron absorption due to GI inflammation. The regions most affected by helminth and malaria (both *Plasmodium falciparum* and *P. vivax*) are sub-Saharan Africa, South Asia and the Middle East, and parts of South America—and both are more prevalent in countries that score lower on the Human Development Index (Battle et al. 2019; Weiss et al. 2019; Hotez and Herricks 2015). Thus, the "hidden hunger" map and the "parasitic infection" map overlap substantially—with the shared factor of poverty linking them together.

We discussed in previous chapters that humans do not have a reliable physiological mechanism to get rid of large amounts of iron, which is why the body puts limits on iron absorption. However, there is one scenario in which iron is removed, sometimes in substantial amounts, in broad swaths of the population. Non-bacterial intestinal infections—helminth worms that feed on human blood via the intestine, removing blood causing small amounts of bleeding that are invisible to the naked eye but nevertheless contribute to iron loss. Helminth worms frequently spend part of their lifespan outside the body in the environment, poised to infect the unsuspecting. For example, hookworm larvae live in the soil, infecting humans via skin (usually through bare feet). They enter the blood stream and leave via the lungs, where they are coughed up and swallowed into the intestine, where they latch on and feed on blood, becoming adults who lay eggs. The eggs are deposited back into the soil to hatch and become larvae, eventually infecting another human host. Other helminths, such as whipworms, roundworms, threadworms, and schistosomes, also put their hosts at risk for anemia, but hookworms carry the greatest risk (Awasthi and Bundy 2007). In some regions of the world, parasite infection is difficult to avoid. The estimates for helminth infection globally range from hundreds of millions to billions of affected individuals.

Parasite load, or the number of helminths that a person has, can vary from person to person and place to place. In general, the intensity of infection is correlated with the number of parasites. The factors that affect the intensity of hookworm infection are poverty, overcrowding, inadequate access to safe water, sanitation facilities, and hygiene, and living in tropical climate and low altitude locations. Parasite load is associated with higher prevalence of anemia, particularly in pregnant women (Larocque et al. 2005). An estimated 24% of the world's population has a helminth infection (World Health Organization 2022b), meaning that the loss of iron from these infections is a substantial force in the iron nutrientscape.

Malaria is another widespread infectious parasite that is associated with iron deficiency. While malaria is not a GI illness, it does infect and destroy red blood cells, meaning that it contributes to anemia. In fact, it is a major cause of anemia in areas where it is endemic. Malaria has a profound negative effect on pregnant women, with maternal anemia being one of the poor outcomes associated with malaria infection. It's estimated that malaria is the third most common cause of maternal death during pregnancy in Africa and is responsible for 400,000 cases of maternal anemia annually (World Health Organization 2022a). Because malaria is transmitted to humans via mosquitos, controlling mosquitos' access to humans is essential for preventing malaria infections. Using mosquito netting around beds, treating standing water, using insecticide, and improving sanitation and nutritional status are all ways to improve malaria infection. Living in areas that are crowded, have standing water, or in housing that does not protect against mosquitoes are all risk factors for the spread of malaria. As is, of course, being poor.

Anthropologists suspect that humans have not always been burdened with such a high degree of helminthic and malaria parasites. Prior to the development of agriculture, most anthropologists argue that humans lived in low-density hunter-gatherer groups whose population sizes were restricted by the carrying capacity of the land. Carrying capacity is just a fancy word for the ability of the environment to support the humans on it. Since humans are so large and have such energy-expending big brains, we need a lot of land to hunt and gather on. Since human groups were less dense, as the hypothesis goes, the ability of pathogens like helminths to infect a person was lower—smaller groups and bigger land use meant that the likelihood of encountering an unsanitary area was lower. That's not to say that these infections were absent in pre-agricultural populations, but that they were less intense and less likely to affect human body iron. Malaria (as well as sickle cell anemia) is also believed to have evolved to high frequencies along with agriculture, which allowed humans to live at higher population densities. This also made it possible for the pathogen to transmit itself more effectively from host to host via its mosquito vector. Today, impoverished conditions are often associated with crowding, along with lack of ability to alleviate its effects. This allows helminths and malaria to flourish, particularly among those who lack easy access to dietary iron.

Poverty is a clear structural barrier to the control of iron-depleting parasites. The inability to access clean water or modern sanitation is a problem that affects billions of people on Earth. The Centers for Disease Control and

Prevention estimates that 1.9 billion people do not have home access to safe water, 2.3 billion do not have home access to safe sanitation, and 2.3 billion do not have adequate hygiene in the home, including access to soap and water (Centers for Disease Control and Prevention 2022). Unfortunately, the seemingly easy solution—give everyone cheap, effective anti-parasite medication—appears to have no overall effect on anemia prevalence or hemoglobin levels (Gyorkos and Gilbert 2014). The more difficult solution—raising the standard of living for the world's poor by creating water and sanitation infrastructure—is costly, time consuming, and difficult to achieve. Still, it works. A systematic review found that providing facilities with adequate waste management reduced the transmission of multiple intestinal helminths (Brooker, Hotez, and Bundy 2008). The real problem is that society allows the poor to live in these circumstances. This, too, is a part of the nutrientscape—the societal conditions that allow people to bleed away their iron is as much a part of human nutritional status as is food.

Structural sexism: Women and hidden hunger

Hidden hungers like iron deficiency disproportionately affect the poor. They also disproportionately affect women, who are responsible for feeding and providing water for their families. Why do women experience more iron deficiency than men? This book has discussed the evolved reproductive vulnerability of women as well as the possibility of culturally bound gendered foods. However, this is not the only explanation for the widespread iron deficiency affecting women—we have also discussed the myth of menstrual iron loss and the ability of homeostasis to recover the iron lost in menstrual blood. That is, given enough dietary iron and well-spaced pregnancies, reproductive-aged women should be able to maintain a physiologically normal level of iron until menopause. The fact that the world's women cannot do so implicates global structures of inequality. While the entities involved in "development"—that is, the increase of economic and industrial infrastructures—have all incorporated language and policies to help the plight of women, the results have fallen far short of the goals.

When the United Nations signed the Millennium Development Goals (MDGs) in 2000, they promised to make gains on eight indicators between the years 1990 and 2015. At least five of these goals are relevant to women's iron status: eliminate extreme poverty and hunger, promote gender equality and empowering women, improve maternal health, combat infectious diseases like malaria, and achieve environmental sustainability (providing access to drinking water and sanitation and improving the lives of slum-dwellers). In the final tally, the ultimate realization of the MDGs was mixed. In all, it is estimated that 21–29 million additional lives were saved beyond prior efforts. Much of this was realized via reductions in tuberculosis, HIV/AIDS, malaria, and child mortality. About 0.5 million maternal deaths were avoided. And an estimated 471 million people were lifted out of extreme poverty (McArthur and Rasmussen 2017). Meanwhile, access to water and sanitation and reduction in undernutrition improved in some countries and declined in others, meaning that the net effect was overall negative. In terms of gender equality, in general the three sub-goals—equality of girls' enrollment in primary school, women's share of paid employment, and women's representation in national parliaments—fared reasonably well, but do not necessarily represent the full spectrum of women's equality, including household decision-making, gender-based violence, time burden of gendered activities, women's property rights, sexual and reproductive rights, and goals to move women out of poverty. The new Sustainable Development Goals make some improvements on that front, with the new goal of "Achieve gender equality and empower all women and girls." This seems like a laudable goal, and much of this is beyond the scope of this book. To keep in line with the goals of this text, I'll begin by asking: What does empowerment mean for women's ability to access iron?

The status of women is a widespread structural barrier within the nutrientscape. Iron, like other resources, does not flow as readily to women. Women's status in many places is unequal; women are less able to access work and if they do, they work for less pay; women are less likely to be educated and are less likely to be able to access household money. To be blunt, women are more likely to live in poverty than men; this happens in both the least wealthy and most wealthy countries like the United States. This realization has led to policy that promotes women's "empowerment," meaning the ability of women to make their own choices, have a sense of self-worth, have access to resources, and to advocate for social change. The mainstream belief about women's empowerment is that by raising the status of women via education and training, women will become empowered within their families and community and gendered inequalities will be reduced.

There is evidence that women's empowerment improves iron status. Most studies of empowering women within the nutrientscape focus on their empowerment in agricultural settings. An example of this kind of empowerment would be owning land that is used to grow food for the household. Empowerment is an interesting concept to operationalize, and most scholars who do so use standardized measurements like the Women's Empowerment in Agriculture Index (WEAI). The WEAI focuses on two aspects of empowerment: women's empowerment across multiple domains of agriculture and the relative gender inequality within the women's household. For example, a study within India found that women were most empowered in agricultural settings where their family owned land to grow food crops compared to women whose families owned cash crops or were landless agricultural workers, and their iron status was better, too (Gupta, Pingali, and Pinstrup-Andersen 2019). In this study, empowerment explained iron status even better than dietary diversity did. This is a stark example of the social nature of the nutrientscape—it is not merely the foods that are available to women, but their ability to access and take ownership of the foods they grow and consume.

Lost in the discussions of women's empowerment is another massive structural barrier: women's reproductive rights. A tremendous gendered structural barrier is that women are often prevented from having the family that they want. A woman's reproductive rights include the right to determine the number and spacing of her births. Reproductive rights are not guaranteed in all places, and contraception is not always accessible or affordable. As I've demonstrated across the pages of this book, a woman's iron status is intimately tied to her reproduction. One assumption is that women want to be able to access contraception and space their births optimally, to best protect their health and the health of their infants. In general, the rule of thumb given by groups like the World Health Organization is "three years between pregnancies, or two years between giving birth and becoming pregnant." They arrived at this conclusion by observing that the lowest risk of both maternal and infant mortality is when mothers space their births three years apart. Any earlier or later, and maternal mortality begins to rise.

This tends to line up with the data I've collected on iron status and return to menstruation, at least in ecologies where diets are lean. Among the Ariaal people of Kenya, women breastfeed their children frequently on demand. I've found that menses resume on average about 18 months after birth. Based on my reconstruction of women's birth histories, the average birth spacing for women in this population is about 28 months, which is just a bit under the recommended 34 months. Given that Ariaal women lose hemoglobin with each successive birth (Chapter 3), their birth spacing could potentially be a bit longer to recoup extra iron with each pregnancy. However, hemoglobin does level off in the population at about 15 months post-birth, meaning that there may be dietary or absorption restrictions after a certain point of iron repletion. Further, women in this community do not have access to contraception or high-quality medical care, and their decision-making about when to have their next child is highly dependent on the wishes of their husbands. Women who feed their infant on demand in the United States, by contrast, resume menses in about 9 months post-birth (Figure 7.1). While I did not collect data about iron status in the U.S. study, it is likely that these women, who were primarily White, middle-to-upper class, and living in an urban area, were mostly able to recoup the iron that they need. Combined with access to contraception and high-quality medical care, it is unsurprising that women in these circumstances would replete iron quickly.

In some places, the assumption that women want to space their births in a certain way or use contraception is not true. Because of the legacies of colonialism and enslavement in many locations around the globe, women in some areas are suspicious of apparent efforts to control their reproduction, rightly suspicious of paternalistic approaches that purport to know the best form of family planning, regardless of a mother's wishes. In the United States, for example, some African American women view attempts at pushing certain types of long-acting contraception as a symbol of reproductive oppression that has been part of the fabric of U.S. slavery and obstetrical care (Reed et al. 2022). Advocating for contraception and birth spacing walks a thin line between advocating for maternal health and oppressing a woman's reproductive rights.

I want to emphasize one final thing. When a woman is disenfranchised in her economy, her community, and her family—her children are, too. Even if—for some reason—you accept that adult women should be iron deficient, a woman's ability to access an iron-rich diet directly impacts her children's iron

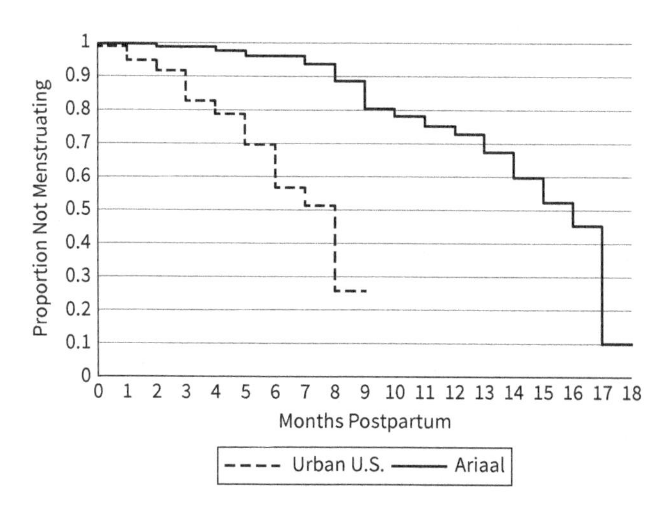

Figure 7.1. Survival curve showing resumption of menstrual periods post-birth in breastfeeding U.S. and Kenyan (Ariaal) women. U.S. women have a steeper curve, indicating a quicker return to menstruation after birth even when exclusively breastfeeding.

status. And consequently, this affects their ability to grow and develop to their fullest potential. A study in Kenya, for example, bears this out: women who are more empowered have children who are less likely to be anemic (Wilunda et al. 2022). This is a reminder to would-be child advocates: if you want to save innocent children, you must also save their mothers.

The nutrientscape within: Hidden hunger, hidden hurts

Hidden hungers like iron deficiency are hidden because the signs are so subtle. When iron deficiency causes anemia, oxygen cannot be effectively carried to the body's tissues, causing fatigue, weakness, pallor of the skin, dizziness, shortness of breath, chest pain, increased heart rate, cold extremities, and headache. Iron deficiency without anemia has a similar set of symptoms: Weakness, fatigue, difficulty in concentrating, and poor work productivity. These symptoms can persist for years without being addressed: after all, isn't everyone tired? Further, screening programs often screen only for anemia (via hemoglobin) and not iron deficiency (via ferritin), so sometime iron deficiency remains hidden. Even though it might seem that these hidden symptoms should affect only the individual—and perhaps even be overcome by someone who can "push through" being tired—their symptoms have wide-ranging consequences for women's participation in households, childcare, and even the market economy. Aside from the structural forces that already keep women from accessing the iron nutrientscape, the symptoms of iron deficiency itself are a barrier to fully experiencing life the way a woman may want.

One of the major hidden consequences of iron deficiency is reduced work capacity and fatigue. No matter what role a woman might perform—physical work, mental work, childcare—iron deficiency impacts her ability to do it. And women are routinely asked to perform labor, even if they are not allowed to participate in market capitalism. Little data exists to document the consequences for reduced work capacity for women outside of exercise science, in part because the nature of the iron deficiency is hidden. What exists paints a picture of economic productivity loss that is a small but significant portion of a nation's gross domestic product (GDP, a measure of a country's economic wealth). One analysis found that 0.57% of GDP was lost due to the physical effects of anemia, and 4.05% of GDP was lost due to both the physical and cognitive effects of anemia in countries where the anemia burden was high (Horton and Ross 2003). This is not only found in developing countries: a historical analysis of widespread U.S. iron fortification in the 1940s found that wages and school attendance improved in areas that had low iron consumption before the fortification mandate (Niemesh 2015). The physical effects were demonstrated in a study of women who worked as tea pluckers in India (Blakstad et al. 2020). The researchers found that women who were clinically anemic picked less tea than women who were not and received 4% fewer wages over a 3-hour period. They found that each 1 g/L increase in hemoglobin level predicted a 1.6% wage increase, with similar effects seen for iron deficiency. While this study featured women performing physical wage labor in West Bengal, India, there is reason to believe that iron deficiency affects many more women in the workforce.

Of course, iron deficiency should not only be considered a problem if it causes loss of salary. Economic consequences are not merely restricted to individual wage labor. For example, the estimated costs of medical care due to anemia are quite high. In Switzerland, for example the costs of medical care for iron deficiency were estimated at 78 million Swiss francs—along with tens of millions in lost human capital due to iron deficiency–related sick leave (Blank et al. 2019).

Further, there are a host of things that women may want to do that could be impaired by iron deficiency, which the cognitive effects alone might impair. Non-market labor, such as domestic labor, suffers too. Childcare especially takes considerable effort. Women may not be able to care for children or other family members as they might want, due to the hidden burden of anemia. Anemia may also impact childcare in even more subtle ways—potentially via mothers' ability to engage with her family in the way she would like. Unfortunately, the effect of maternal anemia on childcare is vastly understudied. Haas and Brownlie (2001 p. 687S) write:

> Even if the iron deficiency does not result in a reduced amount of work performed, the higher cost of performing that work leaves the iron-deficient person less able to engage fully in nonworkplace responsibilities, such as child care, household maintenance, and participation in social and leisure activities. No research has addressed these far-reaching implications of the well-documented effects of iron deficiency anemia (IDA) on work capacity.

While this study noted that future work should involve "extending the analysis of the effect of anemia into nonmarket work activity, such as childcare, household maintenance, and social and leisure activities" (Haas and Brownlie 2001), very little research has actually been done to explore these issues. The fact that there is so little work on the effects of iron deficiency anemia on mothers' social experiences is disappointing, but unsurprising.

Many of the consequences of iron deficiency are shockingly hidden in the brain. There is overwhelming evidence of poorer school performance and hindered cognitive ability in iron-deficient children. The cognitive effects are similar for adults, but the public health impact is thought to be less important since school outcomes are a major concern for human development. Further, it was often thought that once the blood-brain barrier closed, iron levels in the brain remained static. We now know that iron does indeed cross the blood-brain barrier, and evidence is growing that women do suffer neuropsychological effects of iron deficiency (Murray-Kolb 2011). Studies have shown that iron deficiency and/or anemia causes attention difficulties and slower and less accurate performance on standard cognitive tasks, which improves with treatment of the iron deficiency (Murray-Kolb and Beard 2007). These effects on neuropsychological symptoms can have major effects on a woman's quality of life.

One small but growing research area between iron deficiency and the brain is iron deficiency's effects on depression. The symptoms of iron deficiency anemia have considerable overlap with the symptoms of depression: women report irritability, apathy, fatigue, depressive symptoms, and difficulty concentrating. Remember, too, that in Chapter 6 we discussed that the illnesses chlorosis and Cachexia Africana—both of which had psychological symptoms as part of their descriptions—and that racialization has effects on iron status and ultimately depressive symptoms in U.S. women. Reproducing women are especially vulnerable—women's iron levels are low after birth due to pregnancy iron depletion. This is exacerbated by the fact that women are vulnerable to postpartum depression. It's only recently, however, that researchers have recognized the postpartum as a vulnerable time for women's iron status and the full set of symptoms that entails.

Researchers are slowly amassing evidence that iron depletion due to pregnancy and risk of postpartum depression are related. One early study, for example, found that while postpartum iron deficient South African women did not seem different than iron sufficient women on a variety of cognitive and mood tests, their symptoms, including depressed mood, improved after their iron deficiency was treated (Beard et al. 2005). Subsequent studies, reviewed in literature reviews and meta-analyses representing countries in North America, Europe, Africa, Asia, and the Middle East, show that anemia and/or iron deficiency is associated with increased risk of postpartum depression (Wassef, Nguyen, and St-André 2019).

A recent systematic review of 27 studies, representing a variety of populations across the world, found that postpartum anemic and/or iron deficient women were more likely to report symptoms of depression (Moya et al.

2022). They also found that anemic women reported significant improvement in fatigue and depression with treatment across three randomized controlled trials. Finally, they found that women who were anemic were more likely to report negative feelings about their children in two of the four studies that reported this information. The authors concluded that there is good evidence linking postpartum anemia and depression, but that the effects of anemia are not clearly linked to the maternal-child relationship—and that more work on this is required, as researchers have neglected it.

This work demonstrates that the structural inequalities that affect iron deficiency lead to wider problems than mere micronutrient deficiency. Iron deficiency can affect the mental state of those who suffer from it. They, in turn, may be less able to participate in home and work life in the way they would like, further entrenching them in poverty. This is another example of how calls for individuals to better themselves through hard work are unfair—hard work is difficult for those suffering from iron deficiency because, well, they are not at optimal health. The cascading effects of iron deficiency on women's children, families, and communities are difficult to quantify, but are likely considerable. The fact that about 10% of women are iron deficient in the United States—a high-income country with the advantage of advanced agricultural technology and food fortification—means that there is still a long way to go before we fully solve the problem of iron deficiency in countries with a high number of impoverished people.

Conclusion

The mosaic of women's iron deficiency is only complete when poverty is included. Poverty's tesserae are scattered throughout the mosaic—no piece of the image is untouched. Women's evolved reproductive vulnerability to iron deficiency is made worse, a thousand times over, by the human invention of social inequality. The nutrientscape, meant to represent the distribution and flows of iron to and from humans across the globe, is facilitated and constrained by human social structure. Poverty is the number one cause of iron deficiency across the world, constraining women's life choices, including their ability to access a healthy diet and to attain full reproductive justice. Iron deficiency and poverty are a deadly combination, with effects ranging from depression to maternal mortality. Indeed, the effects of iron deficiency may help keep women in poverty. And, when a woman cannot access the iron nutrientscape the way she needs to—not through any fault of her own, but through the failure of the social system—her children are also likely to suffer.

You might walk away from this chapter thinking that better nutritional technology, distributed to impoverished areas, is the solution to women's iron deficiency. This belief, as described by the sociologist Aya Hirata Kimura, is called nutritionism: the belief that primary goal of food is to deliver nutrients (Kimura 2013). Nutritionism is primarily the domain of scientific experts, who are able to both diagnose the problem of unseen, hidden hungers and propose technical solutions to solve them. The "micronutrient turn" Kimura describes as ascendant in the 1990s was associated with a turn toward applying technology to solve hidden hunger and to calculate the human suffering of iron deficiency in economic terms: for example, for every dollar spent on fortification, even more will be gained via economic productivity. Kimura proposes that this is a form of biopower, a term proposed by Foucault to describe how power is exerted over the welfare of individuals and populations, seeking to control, optimize, and subjugate bodies using public health practices. Nutritionism, then, is the exertion of biopower via the deployment of nutrients based on the guidance of experts in nutritional sciences. Put this way, the use of nutritional technology is not a method for the liberation of the world's women, but instead a tool to control them.

So, what is the solution? It's hard to untangle the threads of nutrition, technology, state power, and poverty from one another and to help women who are suffering right now, while still recognizing their agency to choose their own family size and to choose the foods they want to eat. While iron fortification seems to be able to solve the problem of iron deficiency, it means that foods for the poor are extremely processed, often using high amounts of sodium, fat, and simple carbohydrates. In this way, solving one problem (iron deficiency) might lead to another (overnutrition and obesity). Empowerment is a step forward but is an amorphous term that encompasses a wide spectrum of inequalities on multiple levels. It also requires examination of deeply ingrained social structures and cultural beliefs that make women second-class citizens in their communities. The implementation of clean water, hygiene, and sanitation in areas where iron deficiency rates are high would make a large positive difference in the rates of iron deficiency. And finally, full reproductive justice for women, so they can feel free to choose family size without coercion and can access high-quality, culturally appropriate medical care, would go a long way to reducing maternal mortality due to iron deficiency. Since these goals may still have traces of paternalism, colonialism, and inequalities embedded in them, even more radical goals—such as the development of new food systems based in food sovereignty—should be a priority for global women. A nutrientscape emancipated from power structures should be a human right.

8
Conclusions

We see that the mosaic is full of women and their stories. Some are hungry. Some are pregnant. And some are dying. The mosaic is so full, it is difficult to see the whole picture. Squinting, we begin to see the black shadows haunting many of the women. Initially, it seems like these shadows represent a failure of women's biology to manage the stressors of birth or of skipping a meal. But a closer look shows these shadows represent social forces that shape women's stories and constrain what is possible in their lives. Some shadows represent sexism, others racism. The biggest shadow of them all, though, is poverty. Desperately, we begin to look for the light in the mosaic—is there any hope for the women?

Women's evolved biology is not to blame

Biomedicine has a long history of misdeeds when it comes to women's health. Gynecological experimentation on enslaved women, spreading childbed fever from woman to woman in hospital wards on unwashed hands, drugging and restraining women giving birth, and disenfranchising women with traditional knowledge of pregnancy, birth, and breastfeeding are just a few examples of the patriarchal and violent approaches physicians have taken when "helping" women with their reproduction. Since much of our information about the workings of the human body has come from physician-scientists, their biases have led to the widespread belief that women's bodies are degenerative, passive, weak, and inferior. This has shaped our understanding of women's reproduction. Since iron is intimately tied to women's reproduction, anemia has been used as an example of women's inferiority.

Menstruation is the first aspect of women's biology that is criticized and held responsible for women's low iron and propensity for anemia. The immediate targeting of periods, rather than a careful consideration of the role of evolved homeostatic mechanisms in recovering iron levels after the small iron loss of menstruation, can be directly attributed to the bias against women's reproductive biology. The fact that homeostasis was not even studied as a possible compensatory mechanism until very recently demonstrates the depth

Thicker Than Water. Elizabeth M. Miller, Oxford University Press.
DOI: 10.1093/oso/9780197665718.003.0008

and extent of the bias. Only in the past few years have medical authorities in the United States decided that causes of anemia in women other than their periods should even be investigated, meaning that an unknown number of women have suffered from both anemia and a gastrointestinal illness. This demonstrates a callous disregard for women's suffering. It also represents a missed opportunity for decades of research—why not study the role of iron metabolism as a homeostatic system like many other bodily systems? The assumption that menstruation is a consistently depleting force is a modern-day horror story, with draining blood leaving women's bodies weak, anemic shells.

Pregnancy is also blamed for women's low iron, admittedly with good reason. Fetuses absolutely deplete iron at a much higher rate than menstruation. At least some anemia is expected during the late stages of pregnancy due to the depletion of iron stores and the diluting effect of increased maternal bodily fluid. But, as I argued in Chapter 3, reproductive aged women's lower iron levels compared to those of men or post-menopausal women is likely an adaptation to protect embryos and fetuses from the damaging effects of iron. While offspring need iron to develop, they face the same Goldilocks principle as any other living thing—they need the amount of iron to be just right. Since offspring are at the mercy of their mothers' metabolism for their iron needs, a mother's iron set point evolved to moderate between her somatic need for iron, her offspring's need for iron, and the higher risk of injury that iron can pose to developing offspring. This, is, of course, all mediated by homeostatic mechanisms that evolved to maintain balance—as best they can—in the face of inconsistent dietary iron intake.

Homeostasis is a key concept that has been missing from research on women's iron status. I have demonstrated throughout this book that evolution selected to for homeostatic mechanisms that recover iron under the typical conditions of human evolution. Women's physiologies have evolved mechanisms that protect their developing offspring against the harms of iron but leave women vulnerable to low dietary iron. The social conditions from which iron flows to women, the social structures that maintain an impoverished underclass, strain the abilities of homeostasis to maintain appropriate levels of iron. Even biological conditions like inflammation, which alter the course of homeostasis, frequently have a social cause—poor sanitation, crowding, cheap processed foods, and obesity are all associated with both inflammation and poverty. The mechanisms that evolved to keep women's bodies safe from iron fail them in unequal social circumstances.

If evolution is not to blame, then society is. Those who hold power in society do not like disruptions to the status quo, meaning that it is easier to keep blaming women's periods for anemia than it is to admit that societal structures, and not biology, place women in inferior positions. Women's social status has determined that their evolved bodies cannot operate as intended. Biology merely holds up a mirror to society. It is showing us exactly how the worldwide order is failing women. Are we able to see what is needed?

Ensuring women's rights

There is a famous parable that is told in public health circles. It is called the "upstream" or "river" parable.[1] In it, the protagonist—a stand-in for a physician—hears the cry of a person drowning in the river. The protagonist jumps in, fighting against a raging current, drags the person out, and resuscitates them. They hear another cry, and another, each time rescuing the poor drowning soul. It occurs to the protagonist, nearing exhaustion, that maybe they should see who is pushing them in the river.

I have heard versions of this story in which the public health officer, concerned about risk factors, is the one telling the physician about proximate factors that make the river dangerous, a slippery path or perhaps an evil man twirling a mustache loitering about the edge. In my favorite elaboration of this story, there is always an anthropologist standing even farther up the river shouting at the physician and the public health officer. The anthropologist sees that the town's poor are being displaced from the verdant, hilly land away from the river, and are being pushed toward crowded slums lining the river's edge. The displaced poor live so close that they have a tough time avoiding it. Even if they wanted to stay away, they must traverse a dangerous road right next to the river to get work and food. These conditions, of course, create temptation for our mustachioed man, who sneaks up on the unsuspecting. In my version, there is an extra detail—the physician and the public health officers speak a different language. Even if they see the anthropologist shouting, they cannot understand what they are saying. And even if they understood, they are powerless to stop the displacement of the poor. Thus is the frustrating fate of many anthropological endeavors.

In this parable, those pulling iron deficient women out of the river are the healthcare workers who are desperately keeping birthing women and their infants safe and healthy. It is hard to look upstream when faced with the possibility of a woman dying right before your eyes. Upstream, there is both an easy and a difficult way to solve iron deficiency. The "easy" way—which is not easy at all—is to maintain the social status quo and dump more iron into existing food systems, with the hope that it will flow to the people who need it. This

[1] This story is attributed to Irving Zola and recounted in McKinlay (1975).

is the current path taken by those tasked with solving hunger. Fortification of processed foods, biofortification of cereal crops, and supplementation of pregnant women are all steps that public health officials are taking to push iron into the foods of those that need it. However, we have seen that there are multiple drawbacks to this approach: processed foods and cereal grains are not nutritionally dense, and some are exceptionally high in simple carbohydrates and saturated fats, making them a poor choice for everyday meals. In fact, this can give rise to another problem—chronic metabolic disorders due to a low quality, high refined carbohydrate, high fat diet. Supplements, if given in excess, can cause nausea and vomiting in women and can lead to the rejection of the supplementation. These solutions may end up causing as many problems as they attempt to solve; and in any case, widespread fortification of foods has not led to the eradication of iron deficiency, even in the few places where fortification has been executed according to best practices.

The "difficult" way to solve iron deficiency is to look farther upstream to the social inequalities. We know that women are vulnerable to low iron due to the coevolution of iron withholding mechanisms and internal gestation, ensuring that embryos and young fetuses are not harmed by excessive iron. We also know that women are socially vulnerable, less able to access food, resources, income, and capital that would allow them to better their lives and the lives of their children. The difficult solution is to elevate and safeguard women's place in society: protecting a woman's right to reproduce in a healthy, safe, and consensual way and safeguarding her right to grow and/or access nutritious, culturally important foods—foods that can meet her nutritional needs. And women also need governments and non-governmental organizations to address the local and global inequalities that have created a vast underclass of people. No problem, right?

Global organizations have asserted women's reproductive and sexual rights, but these rights are not universally recognized. Unsurprisingly, the poorest women have the least ability to decide how often and with whom they will have children. The right to be able to freely plan your family is critical when addressing women's iron deficiency, since a woman's reproduction is central to her iron status. Women who cannot replete enough iron after their pregnancies during the postpartum will be at risk for iron deficiency during subsequent pregnancies. A woman's postpartum must be vigorously protected—she must have adequate food, rest, and the ability to space her births, whether via breastfeeding or through other means. Her wishes for ideal family size must be respected, and she must not be coerced by others when making decisions about family planning, but instead supported in creating the healthiest outcomes for herself. This includes the right to access supportive medical care across the entire peripartum and to be able to attend well-woman visits when she is not pregnant. Screening for iron deficiency, not just anemia, should be a critical part of women's healthcare, even before a woman becomes pregnant.

The right to sufficient, nutritious, culturally appropriate foods is another upstream piece of the iron deficiency puzzle. While our upstream public health worker sees the need for highly processed, fortified foods, our even further upstream anthropologist can see the unintended consequences of this path: an increasing reliance on capitalist food economies, expensive high tech agricultural techniques, and highly processed, low-quality foods is a formula for disaster. This path leaves women dependent on the whims of market economies and what they will or will not provide. Solutions come from supporting local food systems, putting the means of food production into the hands of the community rather than placing the poor under the control of global foods systems they cannot have any agency over. Improving women's ability to make nutritious food decisions for the household, too, will have the effect of improving the nutritional status of children. And finally, examining the cultural contexts of food production, preparation, and consumption will help us understand the nature of gendered foods and how the underlying beliefs help and hurt women's iron status. Using a framework that helps us understand global and local flows of iron has the potential to increase our understanding of how nutrients move throughout social spaces.

At the heart of these issues are vast global inequalities that have been created by the global community. The history and present of the social structures within and between nations, advantaging some people and places and enormously disadvantaging others, is the main driver of local and global inequalities. The inequalities that lead to increased crowding, lack of sanitation, or lack of ability to access clean water are all major contributors to poor iron status via inflammatory pathways, as just one example. There are no easy solutions to this problem, and certainly a human biologist like me is not qualified to suggest how equality can be achieved in every human context. However, social justice must be the center of any intervention that seeks to alleviate iron deficiency. Hidden hungers cannot fully be addressed without walking upstream and seeing everything—the entire, complete picture—that is pushing women into the river.

Embodiment: The missing link

We are only beginning to understand how social experiences "get under the skin" to alter women's iron status. The obvious pathway of embodiment is diet—how dietary iron flows to and away from individuals depending on multiple factors: gender, race, class, nationality, and more. However, there is still a major black box between "dietary iron" and "iron status." A host of complex, intersecting layers lie between a person's political-economic circumstances and their iron status, meaning that there are multiple ways to study embodiment and to ultimately help reduce the burden of iron deficiency for the world's women.

The study of iron deficiency among women is crying out for better work on embodiment—since nearly all disparities are social, the causes of low iron are social, too. Linking the social—using ethnographic and phenomenological approaches—to the physiological mechanisms governing iron absorption, storage, and reproductive-related iron loss is a critical piece that is missing from research on women's iron deficiency. We need more work on the contexts of nutrient flows to and from women and the experiences that shape their iron intake—particularly social experiences that underly inequalities, such as those due to sexism, racism, and poverty.

We also need a huge amount of work on the absorption of iron in the gut and the role of the microbiome. The microbiome is a clear mediator of iron absorption and learning more about it will explain much of the variation in iron status, as well as offer a potential explanation for the puzzling co-relationship between pica/geophagy and iron deficiency. Since almost all members of the adult gut microbiome need iron too, research is desperately needed to see how gastrointestinal microbes react to dietary iron, how they change the gut environment, and how this alters the homeostasis of iron absorption for their human hosts.

Linking large-scale political economic social structures →social contexts and experiences →diet →the gut environment →iron status →reproduction is a critical pathway of embodiment for iron deficiency. It will not be easy to demonstrate these relationships, but without understanding the entanglements within and between the outside world and the inside body, the world's women will continue to experience hidden hunger. Beyond that, there is a need to represent this pathway dialectically, recognizing that embodiment does not just flow from society to biology, but that low iron impacts women's social experiences, too. The most fruitful way to address this relationship is to make note of the behavioral, mood, and cognitive changes that can accompany iron deficiency and to recognize that they can alter a woman's relationships, her social experiences, and her well-being.

The iron-mind connection

Since iron deficiency has mental/cognitive features, embodiment is a two-way street—iron deficiency is caused by society, and it also changes a woman's relationship to her social world. This can keep women from being able to participate in, and experience, life as healthy people. This problem is underexplored by researchers and completely unknown to healthcare practitioners. How much iron deficiency anemia contributes to, or even causes, mental health disorders is an open question.

As discussed in Chapter 6, historical conditions associated with anemia also had a clear mental component as part of their constellation of symptoms. Chlorosis was associated with being wan and lovesick. Cachexia Africana was thought to be caused, in part, by homesickness and the trauma of becoming enslaved and transported across the Atlantic. Today, the fatiguing and cognitive effects of anemia are known, but the associated mood and behavior changes are invisible to most. Research connecting anemia to depression is only now, in the past decade or so, becoming a topic of interest to scientists. And one of the most curious symptoms of iron deficiency, pica, is stigmatized as a mental health issue even in places where it is considered "culturally appropriate" to engage in pica behaviors like geophagy.

The postpartum is an especially vulnerable time for women. Right after birth is when a woman experiences her lowest iron stores, and it is the time when she is most at risk for postpartum depression. This one-two punch of postpartum depression and postpartum iron deficiency has the potential to be severe for post-birth women. This is exacerbated by healthcare practices that do not follow up with mothers much after birth. In the United States, typically mothers have one postpartum appointment, usually about 6 weeks post-birth. In recognition that it is impossible to cover the substantial number of changes that occur post birth, the American College of Obstetricians and Gynecologists have now started to suggest that women should have multiple appointments during the 12 weeks post-birth in recognition of the support women need in their "fourth trimester."[2] There is also widespread recognition of postpartum depression. However, screening for iron deficiency is not typically performed post-birth. This means that if iron deficiency is a contributing factor to postpartum depression, it will not be recognized in most women's cases. This problem is even worse in contexts with low maternal healthcare coverage, meaning that women may not be seen by a healthcare provider at all during birth, let along screened for postpartum depression or iron deficiency. The postpartum is an important period, one that can help or hinder both a woman's mental well-being and her physical readiness for another child (if she wants one).

Cognitive changes and risk of depression are part of iron's embodiment of inequality. The fact that iron deficiency and/or anemia is not

[2] Whether private insurance will cover more than one visit is another matter entirely.

distributed equally among the world's women mean that the poorest and most disenfranchised women bear the mental and physical burden of postpartum iron deficiency. This problem is poorly recognized by healthcare providers. Instead, it is economists who have pointed it out, reducing the problem to a matter of economic productivity. As this book has argued, caring about women's productivity is not enough. Women's reproductive rights, nutritional rights, and right to a standard of living that ensures wellness must be ensured by all. Truly alleviating the burden of iron deficiency is an incredibly social task, but one that urgently needs to be addressed.

A new mosaic

We've seen enough of this mosaic. We turn away, hoping to leave quickly. We're overwhelmed by the stories we have seen and our lack of ability to change anything. As we rush toward the door, we see a new mosaic being constructed. It shows a healthy pregnant woman surrounded by a supportive community, verdant fields, and, yes, a clean, clear river far in the distance. Without thinking, we reach for a handful of tesserae and begin to work on this vibrant biocultural design.

Bibliography

Al-Amer, O. M., A. A. A. Oyouni, M. A. Alshehri, A. Alasmari, O. R. Alzahrani, S. A. S. Aljohani, N. Alasmael, et al. 2021. "Association of SNPs within TMPRSS6 and BMP2 Genes with Iron Deficiency Status in Saudi Arabia." *PLoS One 16* (11): e0257895. https://doi.org/10.1371/journal.pone.0257895.

Alves, C., A. Saleh, and H. Alaofè. 2019. "Iron-Containing Cookware for the Reduction of Iron Deficiency Anemia among Children and Females of Reproductive Age in Low- and Middle-Income Countries: A Systematic Review." *PLoS One 14* (9): e0221094. https://doi.org/10.1371/journal.pone.0221094.

American Society of Hematology. 2022. "Iron-Deficiency Anemia." Accessed October 13, 2022. https://www.hematology.org/education/patients/anemia/iron-deficiency.

Amir, Dorsa, Matthew R. Jordan, and Richard G. Bribiescas. 2016. "A Longitudinal Assessment of Associations between Adolescent Environment, Adversity Perception, and Economic Status on Fertility and Age of Menarche." *PLoS ONE 11* (6): e0155883. https://doi.org/10.1371/journal.pone.0155883.

Angeli, Adeline, Fabrice Lainé, Audrey Lavenu, Martine Ropert, Karine Lacut, Valérie Gissot, Sylvie Sacher-Huvelin, et al. 2016. "Joint Model of Iron and Hepcidin during the Menstrual Cycle in Healthy Women." *The AAPS Journal 18* (2): 490–504.

Annibale, B., E. Lahner, A. Chistolini, C. Gallucci, E. Di Giulio, G. Capurso, O. Luana, B. Monarca, and G. Delle Fave. 2003. "Endoscopic Evaluation of the Upper Gastrointestinal Tract Is Worthwhile in Premenopausal Women with Iron-Deficiency Anaemia Irrespective of Menstrual Flow." *Scandinavian Journal of Gastroenterology 38* (3): 239–245. https://doi.org/10.1080/00365520310000690a.

Appadurai, Arjun. 1990. "Disjuncture and Difference in the Global Cultural Economy." *Theory, Culture & Society 7* (2–3): 295–310. https://doi.org/10.1177/026327690007002017.

Association of American Medical Colleges. 2021. "Table A-17: MCAT and GPAs for Applicants and Matriculants to U.S. MD-Granting Medical Schools by Primary Undergraduate Major, 2021–2022." Accessed October 4, 2022. https://www.aamc.org/media/6061/download?attachment.

Awasthi, S., and D. Bundy. 2007. "Intestinal Nematode Infection and Anaemia in Developing Countries." *BMJ 334* (7603): 1065–1066. https://doi.org/10.1136/bmj.39211.572905.80.

Bachman, Eric, Thomas G. Travison, Shehzad Basaria, Maithili N. Davda, Wen Guo, Michelle Li, John Connor Westfall, et al. 2013. "Testosterone Induces Erythrocytosis via Increased Erythropoietin and Suppressed Hepcidin: Evidence for a New Erythropoietin/Hemoglobin Set Point." *The Journals of Gerontology: Series A 69* (6): 725–735. https://doi.org/10.1093/gerona/glt154.

Badfar, Gholamreza, Masoumeh Shohani, Ali Soleymani, and Milad Azami. 2019. "Maternal Anemia during Pregnancy and Small for Gestational Age: A Systematic Review and Meta-Analysis." *The Journal of Maternal-Fetal & Neonatal Medicine 32* (10): 1728–1734.

Balk, J., J. M. Connorton, Y. Wan, A. Lovegrove, K. L. Moore, C. Uauy, P. A. Sharp, and P. R. Shewry. 2019. "Improving Wheat as a Source of Iron and Zinc for Global Nutrition." *Nutrition Bulletin 44* (1): 53–59. https://doi.org/10.1111/nbu.12361.

Barba-Moreno, Laura, V. Alfaro-Magallanes, F. Calderón, and A. Peinado. 2020. "Systemic Iron Homeostasis in Female Athletes: Hepcidin, Exercise and Sex Influence." *Archivos de Medicina del Deporte 37*: 348–353.

Barba-Moreno, Laura, Víctor M. Alfaro-Magallanes, Xanne A. K. Janse de Jonge, Angel E. Díaz, Rocío Cupeiro, and Ana B. Peinado. 2020. "Hepcidin and Interleukin-6 Responses to Endurance Exercise over the Menstrual Cycle." *European Journal of Sport Science 22* (2): 218–226. https://doi.org/10.1080/17461391.2020.1853816.

Battle, Katherine E., Tim C. D. Lucas, Michele Nguyen, Rosalind E. Howes, Anita K. Nandi, Katherine A. Twohig, Daniel A. Pfeffer, et al. 2019. "Mapping the Global Endemicity and Clinical Burden of *Plasmodium vivax*, 2000–17: A Spatial and Temporal Modelling Study." *The Lancet 394* (10195): 332–343. https://doi.org/10.1016/S0140-6736(19)31096-7.

Bauman, Dvora, Adir Sommer, Dar Noy, Tal Hammer, and Shoshana Revel-Vilk. 2020. "Under-Diagnosed and Under-Treated Menorrhagia in Young Women Produce Suboptimal Performance." *Blood 136* (Supplement 1): 38–38. https://doi.org/10.1182/blood-2020-139354.

Beard, John L. 2000. "Effectiveness and Strategies of Iron Supplementation during Pregnancy." *The American Journal of Clinical Nutrition 71* (5): 1288S–1294S.

Beard, John L. 2001. "Iron Biology in Immune Function, Muscle Metabolism and Neuronal Functioning." *The Journal of Nutrition 131* (2): 568S–580S.

Beard, John L., Michael K. Hendricks, Eva M. Perez, Laura E. Murray-Kolb, Astrid Berg, Lynne Vernon-Feagans, James Irlam, et al. 2005. "Maternal Iron Deficiency Anemia Affects Postpartum Emotions and Cognition." *The Journal of Nutrition 135* (2): 267–272. https://doi.org/10.1093/jn/135.2.267.

Beutler, Ernest, and Jill Waalen. 2006. "The Definition of Anemia: What Is the Lower Limit of Normal of the Blood Hemoglobin Concentration?" *Blood 107* (5): 1747–1750.

Blackwell, Antoinette Louisa Brown. 1875. *The Sexes throughout Nature*. New York: G.P. Putnam.

Blakstad, M. M., J. E. H. Nevins, S. Venkatramanan, E. M. Przybyszewski, and J. D. Haas. 2020. "Iron Status Is Associated with Worker Productivity, Independent of Physical Effort in Indian Tea Estate Workers." *Applied Physiology, Nutrition, and Metabolism 45* (12): 1360–1367. https://doi.org/10.1139/apnm-2020-0001.

Blank, P. R., Y. Tomonaga, T. D. Szucs, and M. Schwenkglenks. 2019. "Economic Burden of Symptomatic Iron Deficiency—a Survey among Swiss Women." *BMC Women's Health 19* (1): 39. https://doi.org/10.1186/s12905-019-0733-2.

Bloomfield, Arthur L. 1932. "Relations between Primary Hypochromic Anemia and Chlorosis." *Archives of Internal Medicine 50* (2): 328–337.

Boas, Franz. 1912. "Changes in the Bodily Form of Descendants of Immigrants." *American Anthropologist 14* (3): 530–562.

Bodnar, Lisa M., Mary E. Cogswell, and Thad McDonald. 2005. "Have We Forgotten the Significance of Postpartum Iron Deficiency?" *American Journal of Obstetrics and Gynecology 193* (1): 36–44. https://doi.org/10.1016/j.ajog.2004.12.009.

Bozzini, Claudia, Domenico Girelli, Oliviero Olivieri, Nicola Martinelli, Antonella Bassi, Giovanna De Matteis, Ilaria Tenuti, et al. 2005. "Prevalence of Body Iron Excess in the Metabolic Syndrome." *Diabetes Care 28* (8): 2061–2063. https://doi.org/10.2337/diacare.28.8.2061.

Brickley, Megan B. 2018. "Cribra Orbitalia and Porotic Hyperostosis: A Biological Approach to Diagnosis." *American Journal of Physical Anthropology 167* (4): 896–902. https://doi.org/10.1002/ajpa.23701.

Briggs, D. A., D. J. Sharp, D. Miller, and R. G. Gosden. 1999. "Transferrin in the Developing Ovarian Follicle: Evidence for De-novo Expression by Granulosa Cells." *MHR: Basic Science of Reproductive Medicine 5* (12): 1107–1114.

Brooker, Simon, Peter J. Hotez, and Donald A. P. Bundy. 2008. "Hookworm-Related Anaemia among Pregnant Women: A Systematic Review." *PLoS Neglected Tropical Diseases 2* (9): e291.

Brumberg, Joan Jacobs. 1982. "Chlorotic Girls, 1870–1920: A Historical Perspective on Female Adolescence." *Child Development 53* (6): 1468–1477. https://doi.org/10.2307/1130073.

Buckley, Thomas, and Alma Gottlieb. 1988. *Blood Magic*. Berkeley: University of California Press.

Bullough, Vern, and Martha Voght. 2012. "Women, Menstruation, and Nineteenth-Century Medicine." *Bulletin of the History of Medicine 47* (1): 66–82.

Burke, Stacie, and Thomas Duffy. 2022. "Famine, Tea, and Bread in Ireland: C282Y and Modern Human Microevolution." *American Journal of Biological Anthropology 178* (S74): 211–229. https://doi.org/10.1002/ajpa.24520.

Canali, Susanna, Chia-Yu Wang, Kimberly B. Zumbrennen-Bullough, Abraham Bayer, and Jodie L. Babitt. 2017. "Bone Morphogenetic Protein 2 Controls Iron Homeostasis in Mice Independent of BMP6." *American Journal of Hematology 92* (11): 1204–1213. https://doi.org/10.1002/ajh.24888.

Carter, Dan, Yaacov Maor, Simon Bar-Meir, and Benjamin Avidan. 2008. "Prevalence and Predictive Signs for Gastrointestinal Lesions in Premenopausal Women with Iron Deficiency Anemia." *Digestive Diseases and Sciences 53* (12): 3138–3144. https://doi.org/10.1007/s10620-008-0298-7.

Centers for Disease Control and Prevention. 2022. "Global WASH: Fast Facts." Accessed October 6, 2022. https://www.cdc.gov/healthywater/global/wash_statistics.html.

Chao, Kuo-Ching, Chun-Chao Chang, Hung-Yi Chiou, and Jung-Su Chang. 2015. "Serum Ferritin Is Inversely Correlated with Testosterone in Boys and Young Male Adolescents: A Cross-Sectional Study in Taiwan." *PLoS ONE 10* (12): e0144238. https://doi.org/10.1371/journal.pone.0144238.

Chavarro, Jorge E., Janet W. Rich-Edwards, Bernard A. Rosner, and Walter C. Willett. 2006. "Iron Intake and Risk of Ovulatory Infertility." *Obstetrics & Gynecology 108* (5): 1145–1152.

Clancy, Kathryn B. H. 2009. "Reproductive Ecology and the Endometrium: Physiology, Variation, and New Directions." *American Journal of Physical Anthropology 140* (S49): 137–154.

Clancy, Kathryn B. H. 2011. "Iron-Deficiency Is Not Something You Get Just for Being a Lady." Context and Variation (blog). *Scientific American*. https://blogs.scientificamerican.com/context-and-variation/httpblogsscientificamericancomcontext-and-variation20110727iron-deficiency-anemia/.

Clancy, Kathryn B. H., Ilona Nenko, and Grazyna Jasienska. 2006. "Menstruation Does Not Cause Anemia: Endometrial Thickness Correlates Positively with Erythrocyte Count and Hemoglobin Concentration in Premenopausal Women." *American Journal of Human Biology 18* (5): 710–713.

Cogswell, Mary E., Laura Kettel-Khan, and Usha Ramakrishnan. 2003. "Iron Supplement Use among Women in the United States: Science, Policy and Practice." *The Journal of Nutrition 133* (6): 1974S–1977S. https://doi.org/10.1093/jn/133.6.1974S.

Cole, Garrard, and Tony Waldron. 2019. "Cribra Orbitalia: Dissecting an Ill-defined Phenomenon." *International Journal of Osteoarchaeology 29* (4): 613–621.

Connolly, E. L., and M. Guerinot. 2002. "Iron Stress in Plants." *Genome Biology 3* (8): Reviews1024.1. https://doi.org/10.1186/gb-2002-3-8-reviews1024.

Coviello, Andrea D., Beth Kaplan, Kishore M. Lakshman, Tai Chen, Atam B. Singh, and Shalender Bhasin. 2008. "Effects of Graded Doses of Testosterone on Erythropoiesis in Healthy Young and Older Men." *The Journal of Clinical Endocrinology and Metabolism 93* (3): 914–919. https://doi.org/10.1210/jc.2007-1692.

Crosby, William H. 1987. "Whatever Became of Chlorosis?" *JAMA 257* (20): 2799–2800.

Darwin, Charles. 1903. *More Letters of Charles Darwin: A Record of His Work in a Series of Hitherto Unpublished Letters.* Edited by F. Darwin and A Seward. Vol. 2. New York: D. Appleton.

Das, Nupur K., Andrew J. Schwartz, Gabrielle Barthel, Naohiro Inohara, Qing Liu, Amanda Sankar, David R. Hill, et al. 2020. "Microbial Metabolite Signaling Is Required for Systemic Iron Homeostasis." *Cell Metabolism 31* (1): 115–130. https://doi.org/10.1016/j.cmet.2019.10.005.

Datz, C., T. Haas, H. Rinner, F. Sandhofer, W. Patsch, and B. Paulweber. 1998. "Heterozygosity for the C282Y Mutation in the Hemochromatosis Gene Is Associated with Increased Serum Iron, Transferrin Saturation, and Hemoglobin in Young Women: A Protective Role against Iron Deficiency?" *Clinical Chemistry 44* (12): 2429–2432.

Dewey, Kathryn G., and Brietta M Oaks. 2017. "U-shaped Curve for Risk Associated with Maternal Hemoglobin, Iron Status, or Iron Supplementation." *The American Journal of Clinical Nutrition 106* (suppl_6): 1694S–1702S. https://doi.org/10.3945/ajcn.117.156075.

DiGangi, Elizabeth A., and Jonathan D. Bethard. 2021. "Uncloaking a Lost Cause: Decolonizing Ancestry Estimation in the United States." *American Journal of Physical Anthropology 175* (2): 422–436.

Dobzhansky, Theodosius. 1973. "Nothing in Biology Makes Sense Except in the Light of Evolution." *American Biology Teacher 35* (3): 125–129.

Dorsey, Achsah. 2020. "Iron, Infection, and Malnutrition: An Exploration of Childhood Anemia in a Peri-Urban Community in Lima, Peru." PhD diss., University of North Carolina at Chapel Hill.

Dressler, William W. 2020. "Cultural Consensus and Cultural Consonance: Advancing a Cognitive Theory of Culture." *Field Methods 32* (4): 383–398.

Du Bois, William Edward Burghardt. 1906. *The Health and Physique of the Negro American: Report of a Social Study Made under the Direction of Atlanta University, Together with the Proceedings of the Eleventh Conference for the Study of the Negro Problems, Held at Atlanta University, on May the 29th, 1906.* Vol. 11. Atlanta, GA: Atlanta University Press.

Durkheim, Emile. 1963. *Incest: The Nature and Origin of the Taboo.* Translated, with an Introduction, by Edward Sagarin. New York: L. Stuart.

Dutton, Diana B. 1979. "Hematocrit Levels and Race: An Argument against the Adoption of Separate Standards in Screening for Anemia." *Journal of the National Medical Association 71* (10): 945.

Ellison, Peter. 2008. "Energetics, Reproductive Ecology, and Human Evolution." *PaleoAnthropology 2008*: 172–200.

Erwin, H. J. M. Kemna, Tjalsma Harold, L. Willems Hans, and W. Swinkels Dorine. 2008. "Hepcidin: From Discovery to Differential Diagnosis." *Haematologica 93* (1): 90–97. https://doi.org/10.3324/haematol.11705.

Escobar-Morreale, Héctor F. 2012. "Iron Metabolism and the Polycystic Ovary Syndrome." *Trends in Endocrinology & Metabolism 23* (10): 509–515.

Farmer, Paul. 2004. "An Anthropology of Structural Violence." *Current Anthropology 45* (3): 305–325.

Fessler, Daniel. 2002. "Reproductive Immunosuppression and Diet: An Evolutionary Perspective on Pregnancy Sickness and Meat Consumption." *Current Anthropology 43* (1): 19–61.

Figlio, Karl. 1978. "Chlorosis and Chronic Disease in Nineteenth-century Britain: The Social Constitution of Somatic Illness in a Capitalist Society." *Social History 3* (2): 167–197.

Fine, Cordelia. 2010. *Delusions of Gender: How Our Minds, Society, and Neurosexism Create Difference.* New York: WW Norton & Company.

Finn, CA. 1994. "The Adaptive Significance of Menstruation: The Meaning of Menstruation." *Human Reproduction 9* (7): 1202–1204.

Fratkin, Elliot, and Eric Abella Roth. 2006. *As Pastoralists Settle: Social, Health, and Economic Consequences of the Pastoral Sedentarization in Marsabit District, Kenya*. New York: Springer Science & Business Media.

Gagnon, David R., Ting-Jie Zhang, Frederick N. Brand, and William B. Kannel. 1994. "Hematocrit and the Risk of Cardiovascular Disease—The Framingham Study: A 34-Year Follow-up." *American Heart Journal 127* (3): 674–682. https://doi.org/https://doi.org/10.1016/0002-8703(94)90679-3.

Garber, Arkadiy I., Kenneth H. Nealson, Akihiro Okamoto, Sean M. McAllister, Clara S. Chan, Roman A. Barco, and Nancy Merino. 2020. "FeGenie: A Comprehensive Tool for the Identification of Iron Genes and Iron Gene Neighborhoods in Genome and Metagenome Assemblies." *Frontiers in Microbiology 67*: 37.

Garn, Stanley M. 1981. "Lower Hematocrit Levels in Blacks Are Not Due to Diet or Socioeconomic Factors." *Pediatrics 67* (4): 580.

Garn, Stanley M., H. A. Shaw, K. E. Guire, and K. McCabe. 1977. "Apportioning Black-White Hemoglobin and Hematocrit Differences during Pregnancy." *The American Journal of Clinical Nutrition 30* (4): 461–462.

Garn, Stanley M., Nathan J. Smith, and Diane C. Clark. 1974. "Race Differences in Hemoglobin Levels." *Ecology of Food and Nutrition 3* (4): 299–301.

Gavin, Norma I., E. Kathleen Adams, Katherine E. Hartmann, M. Beth Benedict, and Monique Chireau. 2004. "Racial and Ethnic Disparities in the Use of Pregnancy-Related Health Care among Medicaid Pregnant Women." *Maternal and Child Health Journal 8* (3): 113–126. https://doi.org/10.1023/B:MACI.0000037645.63379.62.

Geller, Pamela L. 2009. "Bodyscapes, Biology, and Heteronormativity." *American Anthropologist 111* (4): 504–516.

Georgsen, Maja, Maria Christine Krog, Anne-Sofie Korsholm, Helene Westring Hvidman, Astrid Marie Kolte, Andreas Stribolt Rigas, Henrik Ullum, et al. 2021. "Serum Ferritin Level Is Inversely Related to Number Of Previous Pregnancy Losses in Women with Recurrent Pregnancy Loss." *Fertility and Sterility 115* (2): 389–396.

Goodman, Alan. 1996. "Glorification of the Genes: Genetic Determinism and Racism in Science." In *The Life Industry: Biodiversity, People and Profits*, edited by M. Baumann, J. Bell, F. Koechlin and M. Pimbert, 149–160. London: Intermediate Technology Publications.

Goodman, Alan H., and Thomas Leland Leatherman. 1998. *Building a New Biocultural Synthesis: Political-Economic Perspectives on Human Biology*. Ann Arbor: University of Michigan Press.

Gravlee, Clarence C. 2009. "How Race Becomes Biology: Embodiment of Social Inequality." *American Journal of Physical Anthropology 139* (1): 47–57.

Green, B. T., and D. C. Rockey. 2004. "Gastrointestinal Endoscopic Evaluation of Premenopausal Women with Iron Deficiency Anemia." *Journal of Clinical Gastroenterology 38* (2): 104–109. https://doi.org/10.1097/00004836-200402000-00004.

Guo, Yifan, Na Zhang, Daoqiang Zhang, Quanzhong Ren, Tomas Ganz, Sijin Liu, and Elizabeta Nemeth. 2019. "Iron Homeostasis in Pregnancy and Spontaneous Abortion." *American Journal of Hematology 94* (2): 184–188.

Gupta, Soumya, Prabhu Pingali, and Per Pinstrup-Andersen. 2019. "Women's Empowerment and Nutrition Status: The Case of Iron Deficiency in India." *Food Policy 88*: 101763. https://doi.org/10.1016/j.foodpol.2019.101763.

Gyorkos, Theresa W., and Nicolas L. Gilbert. 2014. "Blood Drain: Soil-Transmitted Helminths and Anemia in Pregnant Women." *PLOS Neglected Tropical Diseases 8* (7): e2912. https://doi.org/10.1371/journal.pntd.0002912.

Haas, Jere D., and Thomas Brownlie, IV. 2001. "Iron Deficiency and Reduced Work Capacity: A Critical Review of the Research to Determine a Causal Relationship." *The Journal of Nutrition 131* (2): 676S–690S. https://doi.org/10.1093/jn/131.2.676S.

Haig, D. 1993. "Genetic Conflicts in Human Pregnancy." *Quarterly Review of Biology 68* (4): 495–532. https://doi.org/10.1086/418300.

Hallberg, L., and L. Rossander-Hulten. 1991. "Iron Requirements in Menstruating Women." *The American Journal of Clinical Nutrition 54* (6): 1047–1058.

Haller, John S. 1972. "The Negro and the Southern Physician: A Study of Medical and Racial Attitudes 1800–1860." *Medical History 16* (3): 238–253.

Harrison, Faye V. 2016. "Engaging Theory in the New Millennium." In *The Routledge Companion to Contemporary Anthropology*, edited by Simon Coleman, Susan Hyatt, Ann Kingsolver, 43–72. New York: Routledge.

Heath, K. M., J. H. Axton, J. M. McCullough, and N. Harris. 2016. "The Evolutionary Adaptation of the C282Y Mutation to Culture and Climate during the European Neolithic." *American Journal of Physical Anthropology 160* (1): 86–101. https://doi.org/10.1002/ajpa.22937.

Hodgins, S., and A. D'Agostino. 2014. "The Quality-Coverage Gap in Antenatal Care: Toward Better Measurement of Effective Coverage." *Global Health in Science and Practice 2* (2): 173–181. https://doi.org/10.9745/ghsp-d-13-00176.

Horton, Sue, and Carol Levin. 2001. Commentary on "Evidence That Iron Deficiency Anemia Causes Reduced Work Capacity." *The Journal of Nutrition 131* (2): 691S–696S. https://doi.org/10.1093/jn/131.2.691S.

Hogarth, Rana A. 2017. *Medicalizing Blackness: Making Racial Difference in the Atlantic World, 1780–1840*. Chapel Hill, NC: UNC Press Books.

Holland, Heinrich D. 2006. "The Oxygenation of the Atmosphere and Oceans." *Philosophical Transactions of the Royal Society B: Biological Sciences 361* (1470): 903–915. https://doi.org/doi:10.1098/rstb.2006.1838.

Hollerer, I., A. Bachmann, and M. U. Muckenthaler. 2017. "Pathophysiological Consequences and Benefits of HFE Mutations: 20 Years of Research." *Haematologica 102* (5): 809–817. https://doi.org/10.3324/haematol.2016.160432.

Horton, Susan, and Jay Ross. 2003. "The Economics of Iron Deficiency." *Food Policy 28* (1): 51–75.

Hotez, Peter J., and Jennifer R. Herricks. 2015. "Helminth Elimination in the Pursuit of Sustainable Development Goals: A 'Worm Index' for Human Development." *PLoS Neglected Tropical Diseases 9* (4): e0003618. https://doi.org/10.1371/journal.pntd.0003618.

Hou, Yanli, Shuping Zhang, Lei Wang, Junping Li, Guangbo Qu, Jiuyang He, Haiqin Rong, Hong Ji, and Sijin Liu. 2012. "Estrogen Regulates Iron Homeostasis through Governing Hepatic Hepcidin Expression Via an Estrogen Response Element." *Gene 511* (2): 398–403.

Hubbard, A. C., S. Bandyopadhyay, B. S. Wojczyk, S. L. Spitalnik, E. A. Hod, and K. A. Prestia. 2013. "Effect of Dietary Iron on Fetal Growth in Pregnant Mice." *Comparative Medicine 63* (2): 127–135.

Hueston, W, and A McLeod. 2012. "Overview of the Global Food System: Changes over Time/Space and Lessons for Future Food Safety." In *Institute of Medicine (US). Improving Food Safety Through a One Health Approach: Workshop Summary, A5*, rapporteurs, Eileen R. Choffnes, David A. Relman, LeighAnne Olsen, Rebekah Hutton, and Alison Mack, 189–197. Washington, DC: National Academies Press.

Hurrell, R. F. 2021. "Iron Fortification Practices and Implications for Iron Addition to Salt." *Journal of Nutrition 151* (Suppl 1): 3s–14s. https://doi.org/10.1093/jn/nxaa175.

Hurrell, R. F., P. Ranum, S. de Pee, R. Biebinger, L. Hulthen, Q. Johnson, and S. Lynch. 2010. "Revised Recommendations for Iron Fortification of Wheat Flour and an Evaluation of the Expected Impact of Current National Wheat Flour Fortification Programs." *Food and Nutrition Bulletin 31* (1 Suppl): S7–21. https://doi.org/10.1177/15648265100311s102.

Igbinosa, Irogue, Stephanie A Leonard, Alexander J Butwick, and Deirdre J. Lyell. 2020. "464: Antepartum Anemia and Racial/Ethnic Disparities in Blood Transfusion in California." *American Journal of Obstetrics & Gynecology 222* (1): S304.

Ina, Hollerer, Bachmann André, and U. Muckenthaler Martina. 2017. "Pathophysiological Consequences and Benefits of HFE Mutations: 20 Years of Research." *Haematologica 102* (5): 809–817. https://doi.org/10.3324/haematol.2016.160432.

Institute of Medicine (US) Committee on Nutritional Status during Pregnancy and Lactation. 1990. *Nutrition During Pregnancy: Part I Weight Gain: Part II Nutrient Supplements.* Washington, DC: National Academies Press.

Jackson, RT. 1990. "Separate Hemoglobin Standards for Blacks and Whites: A Critical Review of the Case for Separate and Unequal Hemoglobin Standards." *Medical Hypotheses 32* (3): 181–189.

Jacobi, Mary Putnam. 1877. *The Question of Rest for Women during Menstruation.* New York: G.P. Putnam's Sons.

Jaeggi, Tanja, Guus A. M. Kortman, Diego Moretti, Christophe Chassard, Penny Holding, Alexandra Dostal, Jos Boekhorst, et al. 2015. "Iron Fortification Adversely Affects the Gut Microbiome, Increases Pathogen Abundance and Induces Intestinal Inflammation in Kenyan Infants." *Gut 64* (5): 731–742.

Jiang, W., M. Constante, and M. M. Santos. 2008. "Anemia Upregulates Lipocalin 2 in the Liver and Serum." *Blood Cells, Molecules, and Diseases 41* (2): 169–174. https://doi.org/10.1016/j.bcmd.2008.04.006.

Johnson, Cassandra M., Joseph R. Sharkey, Mellanye J. Lackey, Linda S. Adair, Allison E. Aiello, Sarah K. Bowen, Wei Fang, Valerie L. Flax, and Alice S. Ammerman. 2018. "Relationship of Food Insecurity to Women's Dietary Outcomes: A Systematic Review." *Nutrition Reviews 76* (12): 910–928.

Johnson-Spear, Mary A., and Ray Yip. 1994. "Hemoglobin Difference between Black and White Women with Comparable Iron Status: Justification for Race-Specific Anemia Criteria." *The American Journal of Clinical Nutrition 60* (1): 117–121.

Kalisch-Smith, Jacinta I., Nikita Ved, Dorota Szumska, Jacob Munro, Michael Troup, Shelley E. Harris, Helena Rodriguez-Caro, et al. 2021. "Maternal Iron Deficiency Perturbs Embryonic Cardiovascular Development in Mice." *Nature Communications 12* (1): 3447. https://doi.org/10.1038/s41467-021-23660-5.

Karakochuk, C. D., S. Y. Hess, D. Moorthy, S. Namaste, M. E. Parker, A. I. Rappaport, R. Wegmüller, and O. Dary. 2019. "Measurement and Interpretation of Hemoglobin Concentration in Clinical and Field Settings: A Narrative Review." *Annals of the New York Academies of Sciences 1450* (1): 126–146. https://doi.org/10.1111/nyas.14003.

Kavle, J. A., R. J. Stoltzfus, F. Witter, J. M. Tielsch, S. S. Khalfan, and L. E. Caulfield. 2008. "Association between Anaemia during Pregnancy and Blood Loss At and After Delivery among Women with Vaginal Births in Pemba Island, Zanzibar, Tanzania." *Journal of Health and Population Nutrition 26* (2): 232–240.

Kepczyk, T., J. E. Cremins, B. D. Long, M. B. Bachinski, L. R. Smith, and P. R. McNally. 1999. "A Prospective, Multidisciplinary Evaluation Of Premenopausal Women with Iron-Deficiency Anemia." *The American Journal of Gastroenterology 94* (1): 109–115. https://doi.org/10.1111/j.1572-0241.1999.00780.x.

Keynes, M., and T. M. Cox. 2008. "William Bateson, The Rediscoverer of Mendel." *Journal of the Royal Society of Medicine 101* (3): 104. https://doi.org/10.1258/jrsm.2008.081011.

Kimura, Aya Hirata. 2013. *Hidden Hunger: Gender and the Politics of Smarter Foods.* Ithaca, NY: Cornell University Press.

King, Helen. 2004. *The Disease of Virgins: Green Sickness, Chlorosis and the Problems of Puberty.* New York, NY: Routledge.

Kiple, Kenneth F. 2002. *The Caribbean Slave: A Biological History.* Cambridge: Cambridge University Press.

Kluger, Matthew J., Wieslaw Kozak, Carole A. Conn, Lisa R. Leon, and Dariusz Soszynski. 1998. "Role of Fever in Disease." *Annals of the New York Academy of Sciences 856* (1): 224–233.

Ko, Cynthia W., Shazia M. Siddique, Amit Patel, Andrew Harris, Shahnaz Sultan, Osama Altayar, and Yngve Falck-Ytter. 2020. "AGA Clinical Practice Guidelines on the Gastrointestinal Evaluation of Iron Deficiency Anemia." *Gastroenterology 159* (3): 1085–1094.

Kobayashi, Motoi, Tomohiro Suhara, Yuichi Baba, Nicholas K. Kawasaki, Jason K. Higa, and Takashi Matsui. 2018. "Pathological Roles of Iron in Cardiovascular Disease." *Current Drug Targets 19* (9): 1068–1076.

Kortman, Guus A.M., Manuela Raffatellu, Dorine W. Swinkels, and Harold Tjalsma. 2014. "Nutritional Iron Turned Inside Out: Intestinal Stress from a Gut Microbial Perspective." *FEMS Microbiology Reviews 38* (6): 1202–1234. https://doi.org/10.1111/1574-6976.12086.

Krieger, Nancy. 2001. "Theories for Social Epidemiology in the 21st Century: An Ecosocial Perspective." *International Journal of Epidemiology 30* (4): 668–677.

Kuchta, Brigitte. 1982. "Experiments and Ultrastructural Investigations on the Mouse Embryo during Early Teratogen-Sensitive Stages." *Cells Tissues Organs 113* (3): 218–225.

Larocque, Renee, Martin Casapia, Eduardo Gotuzzo, and Theresa W Gyorkos. 2005. "Relationship between Intensity of Soil-Transmitted Helminth Infections and Anemia during Pregnancy." *The American Journal of Tropical Medicine and Hygiene 73* (4): 783–789.

Leatherman, Thomas, and Alan Goodman. 2020. "Building on the Biocultural Syntheses: 20 Years and Still Expanding." *American Journal of Human Biology 32* (4): e23360.

Lee, Eun-Sook, Eun-Ji Song, So-Young Lee, So-Lim Park, Daeyoung Kim, Daniel Kim, Jae-Hwan Kim, Seong-Il Lim, and Young-Do Nam. 2018. "Effects of Bentonite Bgp35b-p on the Gut Microbiota of Mice Fed a High-Fat Diet." *Journal of the Science of Food and Agriculture 98* (11): 4369–4373. https://doi.org/10.1002/jsfa.8934.

Lee, Yong-Jae, and Hong-Bae Kim. 2020. "Association between Anaemia and Adult Depression: A Systematic Review and Meta-Analysis of Observational Studies." *Journal of Epidemiology and Community Health 74* (7): 565–572. https://doi.org/10.1136/jech-2020-213842.

Li, Songlin, Changxing Jiang, Xiaoming Chen, Hui Wang, and Jing Lin. 2014. "*Lactobacillus casei* Immobilized onto Montmorillonite: Survivability in Simulated Gastrointestinal Conditions, Refrigeration and Yogurt." *Food Research International 64*: 822–830.

Li, Xiang, David K. Rhee, Rajeev Malhotra, Claire Mayeur, Liam A. Hurst, Emily Ager, Georgia Shelton, et al. 2016. "Progesterone Receptor Membrane Component-1 Regulates Hepcidin Biosynthesis." *The Journal of Clinical Investigation 126* (1): 389–401. https://doi.org/10.1172/JCI83831.

Li, Yan Qin, Xiao Xiao Cao, Bin Bai, Jing Ni Zhang, Ming Qi Wang, and Yao Hui Zhang. 2014. "Severe Iron Deficiency Is Associated with a Reduced Conception Rate in Female Rats." *Gynecologic and Obstetric Investigation 77* (1): 19–23.

Liu, Han, Congmin Wang, Xueling Gu, Jing Zhao, Cunxi Nie, Wenju Zhang, and Xi Ma. 2020. "Dietary Montmorillonite Improves the Intestinal Mucosal Barrier and Optimizes the Intestinal Microbial Community of Weaned Piglets." *Frontiers in Microbiology 11*: 593056.

Liu, Yongsheng. 2008. "A New Perspective on Darwin's Pangenesis." *Biological Reviews 83* (2): 141–149.

Lock, Margaret M. 1994. *Encounters with Aging: Mythologies of Menopause in Japan and North America*. Berkeley: University of California Press.

Lock, Margaret M., and Vinh-Kim Nguyen. 2018. *An Anthropology of Biomedicine*. New York: John Wiley & Sons.

López-Millán, Ana-Flor, Michael A. Grusak, Anunciación Abadía, and Javier Abadía. 2013. "Iron Deficiency in Plants: An Insight from Proteomic Approaches." *Frontiers in Plant Science 4*: 254.

Loudon, Irvine S. 1980. "Chlorosis, Anaemia, and Anorexia Nervosa." *British Medical Journal 281* (6256): 1669.

Martin, Emily. 1991. "The Egg and the Sperm: How Science Has Constructed a Romance Based on Stereotypical Male-Female Roles." *Signs: Journal of Women in Culture and Society 16* (3): 485–501.

Martin, Emily. 2001. *The Woman in the Body: A Cultural Analysis of Reproduction.* New York: Beacon Press.

Mazure, Carolyn M., and Daniel P Jones. 2015. "Twenty Years and Still Counting: Including Women as Participants and Studying Sex and Gender in Biomedical Research." *BMC Women's Health 15* (1): 1–16.

McArthur, John, and Krista Rasmussen. 2017. "How Successful Were the Millennium Development Goals." *The Guardian*, March 30.

McKinlay, J. B. 1975. "A Case for Refocusing Upstream: The Political Economy Of Illness." In *Patients, Physicians, and Illness: A Sourcebook in Behavioral Science and Health.* 3rd ed., edited by E. G. Jaco, 9–25. New York: Free Press.

MedlinePlus. 2022. "Hereditary Hemochromatosis." *U.S. National Library of Medicine.* Accessed September 20, 2021. https://medlineplus.gov/genetics/condition/hereditary-hemochromatosis/#frequency.

Miller, Elizabeth M. 2010. "Maternal Hemoglobin Depletion in a Settled Northern Kenyan Pastoral Population." *American Journal of Human Biology 22* (6): 768–774.

Miller, Elizabeth M. 2014. "Iron Status and Reproduction in US Women: National Health and Nutrition Examination Survey, 1999-2006." *PLoS One 9* (11): e112216. https://doi.org/10.1371/journal.pone.0112216.

Miller, Elizabeth M. 2016. "The Reproductive Ecology of Iron in Women." *American Journal of Physical Anthropology 159*: 172–195.

Miller, Elizabeth M., and Maie Khalil. 2019. "Iron and Fecundity among Tsimane'women of Bolivia." *Evolution, Medicine, and Public Health 2019* (1): 111–120.

Milman, N., M. Kirchhoff, and T. Jørgensen. 1992. "Iron Status Markers, Serum Ferritin and Hemoglobin in 1359 Danish Women in Relation to Menstruation, Hormonal Contraception, Parity, and Postmenopausal Hormone Treatment." *Annals of Hematology 65* (2): 96–102. https://doi.org/10.1007/BF01698138.

Morgan, Maria A., Britta L Anderson, Hal Lawrence, and Jay Schulkin. 2012. "Well-Woman Care among Obstetrician-Gynecologists: Opportunity for Preconception Care." *The Journal of Maternal-Fetal & Neonatal Medicine 25* (6): 595–599.

Moya, Ernest, Nomsa Phiri, Augustine T. Choko, Martin N. Mwangi, and Kamija S. Phiri. 2022. "Effect of Postpartum Anaemia on Maternal Health-Related Quality of Life: A Systematic Review and Meta-Analysis." *BMC Public Health 22* (1): 364. https://doi.org/10.1186/s12889-022-12710-2.

Muckenthaler, Martina U., Heimo Mairbäurl, and Max Gassmann. 2020. "Iron Metabolism in High-Altitude Residents." *Journal of Applied Physiology 129* (4): 920–925. https://doi.org/10.1152/japplphysiol.00019.2020.

Muleviciene, A., F. D'Amico, S. Turroni, M. Candela, and A. Jankauskiene. 2018. "Iron Deficiency Anemia-Related Gut Microbiota Dysbiosis in Infants and Young Children: A Pilot Study." *Acta Microbiologica Immunologica Hungary 65* (4): 551–564. https://doi.org/10.1556/030.65.2018.045.

Murphy, Kevin M., Philip G. Reeves, and Stephen S. Jones. 2008. "Relationship between Yield and Mineral Nutrient Concentrations in Historical and Modern Spring Wheat Cultivars." *Euphytica 163* (3): 381–390. https://doi.org/10.1007/s10681-008-9681-x.

Murray-Kolb, L. E. 2011. "Iron Status and Neuropsychological Consequences in Women of Reproductive Age: What Do We Know and Where Are We Headed?" *Journal of Nutrition 141* (4): 747s–755s. https://doi.org/10.3945/jn.110.130658.

Murray-Kolb, L. E., and John L. Beard. 2007. "Iron Treatment Normalizes Cognitive Functioning in Young Women." *The American Journal of Clinical Nutrition 85* (3): 778–787. https://doi.org/10.1093/ajcn/85.3.778.

Muthayya, Sumithra, Jee Hyun Rah, Jonathan D Sugimoto, Franz F Roos, Klaus Kraemer, and Robert E Black. 2013. "The Global Hidden Hunger Indices and Maps: An Advocacy Tool for Action." *PloS ONE 8* (6): e67860.

Mutter, Andrew C., Alexei M. Tyryshkin, Ian J. Campbell, Saroj Poudel, George N. Bennett, Jonathan J. Silberg, Vikas Nanda, and Paul G. Falkowski. 2019. "De novo Design of Symmetric Ferredoxins That Shuttle Electrons in vivo." *Proceedings of the National Academy of Sciences 116* (29): 14557–14562. https://doi.org/10.1073/pnas.1905643116.

Mycek, Mari Kate. 2018. "Meatless Meals and Masculinity: How Veg* Men Explain Their Plant-Based Diets." *Food and Foodways 26* (3): 223–245.

Nakashige, T. G., B. Zhang, C. Krebs, and E. M. Nolan. 2015. "Human Calprotectin Is an Iron-Sequestering Host-Defense Protein." *Nature Chemical Biology 11* (10): 765–771. https://doi.org/10.1038/nchembio.1891.

Nguyen, Vinh-Kim, and Karine Peschard. 2003. "Anthropology, Inequality, and Disease: A Review." *Annual Review of Anthropology 32*: 447–474.

Niemesh, Gregory T. 2015. "Ironing Out Deficiencies: Evidence from the United States on the Economic Effects of Iron Deficiency." *The Journal of Human Resources 50* (4): 910–958.

Niewöhner, Jörg, and Margaret Lock. 2018. "Situating Local Biologies: Anthropological Perspectives on Environment/Human Entanglements." *BioSocieties 13* (4): 681–697.

Ortiz-Rivera, Y., R. Sánchez-Vega, N. Gutiérrez-Méndez, J. León-Félix, C. Acosta-Muñiz, and D. R. Sepulveda. 2017. "Production of Reuterin in a Fermented Milk Product by *Lactobacillus reuteri*: Inhibition of Pathogens, Spoilage Microorganisms, and Lactic Acid Bacteria." *Journal of Dairy Science 100* (6): 4258–4268.

Owais, A., C. Merritt, C. Lee, and Z. A. Bhutta. 2021. "Anemia among Women of Reproductive Age: An Overview of Global Burden, Trends, Determinants, and Drivers of Progress in Low- and Middle-Income Countries." *Nutrients 13* (8): 2745. https://doi.org/10.3390/nu13082745.

Owens, Deirdre Cooper. 2017. *Medical Bondage: Race, Gender, and the Origins of American Gynecology*. Athens: University of Georgia Press.

Paganini, D., and M. B. Zimmermann. 2017. "The Effects of Iron Fortification and Supplementation on the Gut Microbiome and Diarrhea in Infants and Children: A Review." *American Journal of Clinical Nutrition 106* (Suppl 6): 1688s–1693s. https://doi.org/10.3945/ajcn.117.156067.

Pahwa, R., A. Goyal, , and I. Jialal. 2021. "Chronic Inflammation." *StatPearls*. StatPearls Publishing. Last Modified 8/11/2021. Accessed September 20, 2021. https://www.ncbi.nlm.nih.gov/books/NBK493173/.

Pasolli, Edoardo, Francesca De Filippis, Italia E. Mauriello, Fabio Cumbo, Aaron M. Walsh, John Leech, Paul D. Cotter, Nicola Segata, and Danilo Ercolini. 2020. "Large-Scale Genome-Wide Analysis Links Lactic Acid Bacteria from Food with the Gut Microbiome." *Nature Communications 11* (1): 2610. https://doi.org/10.1038/s41467-020-16438-8.

Pérez-Guzmán, L., K. R. Bogner, and B. H. Lower. 2010. "Earth's Ferrous Wheel." *Nature Education 3* (10): 32.

Pizzino, Gabriele, Natasha Irrera, Mariapaola Cucinotta, Giovanni Pallio, Federica Mannino, Vincenzo Arcoraci, Francesco Squadrito, Domenica Altavilla, and Alessandra Bitto. 2017. "Oxidative Stress: Harms and Benefits for Human Health." *Oxidative Medicine and Cellular Longevity 2017*: 8416763–8416763. https://doi.org/10.1155/2017/8416763.

Radlowski, E. C., and R. W. Johnson. 2013. "Perinatal Iron Deficiency and Neurocognitive Development." *Frontiers in Human Neuroscience 7*: 585. https://doi.org/10.3389/fnhum.2013.00585.

Rahmati, Shoboo, Milad Azami, Gholamreza Badfar, Naser Parizad, and Kourosh Sayehmiri. 2020. "The Relationship between Maternal Anemia during Pregnancy with Preterm Birth: A Systematic Review and Meta-Analysis." *The Journal of Maternal-Fetal & Neonatal Medicine 33* (15): 2679–2689.

Ratledge, Colin, and Lynn G. Dover. 2000. "Iron Metabolism in Pathogenic Bacteria." *Annual Reviews in Microbiology 54* (1): 881–941.

Reed, Reiley, Olga Osby, Mary Nelums, Cassandra Welchlin, Rassidatou Konate, and Kelsey Holt. 2022. "Contraceptive Dare Experiences and Preferences among Black Women in Mississippi: A Qualitative Study." *Contraception 114*: 18–25. https://doi.org/10.1016/j.contraception.2022.05.009.

Robertson, Michael P., and Gerald F. Joyce. 2012. "The Origins of the RNA World." *Cold Spring Harbor Perspectives in Biology 4* (5): a003608. https://doi.org/10.1101/cshperspect.a003608.

Rook, Graham A. W. 2010. "99th Dahlem Conference on Infection, Inflammation and Chronic Inflammatory Disorders: Darwinian Medicine and the 'Hygiene'or 'Old Friends' Hypothesis." *Clinical & Experimental Immunology 160* (1): 70–79.

Rushton, D. H., I. D. Ramsay, J. J. Gilkes, and M. J. Norris. 1991. "Ferritin and Fertility." *The Lancet 337* (8756): 1554–1554.

Rusu, Ioana Gabriela, Ramona Suharoschi, Dan Cristian Vodnar, Carmen Rodica Pop, Sonia Ancuţa Socaci, Romana Vulturar, Magdalena Istrati, et al. 2020. "Iron Supplementation Influence on the Gut Microbiota and Probiotic Intake Effect in Iron Deficiency—A Literature-Based Review." *Nutrients 12* (7): 1993.

Saez-Lara, M. J., C. Gomez-Llorente, J. Plaza-Diaz, and A. Gil. 2015. "The Role of Probiotic Lactic Acid Bacteria and Bifidobacteria in the Prevention and Treatment of Inflammatory Bowel Disease and Other Related Diseases: A Systematic Review of Randomized Human Clinical Trials." *Biomedical Research International 2015*: 505878. https://doi.org/10.1155/2015/505878.

Sangkhae, Veena, A. L. Fisher, S. Wong, M. D. Koenig, L. Tussing-Humphreys, A. Chu, M. Lelić, T. Ganz, and E. Nemeth. 2020. "Effects of Maternal Iron Status on Placental and Fetal Iron Homeostasis." *Journal of Clinical Investigation 130* (2): 625–640. https://doi.org/10.1172/jci127341.

Sangkhae, Veena, Allison L. Fisher, Kristine J. Chua, Piotr Ruchala, Tomas Ganz, and Elizabeta Nemeth. 2020. "Maternal Hepcidin Determines Embryo Iron Homeostasis in Mice." *Blood 136* (19): 2206–2216. https://doi.org/10.1182/blood.2020005745.

Sangkhae, Veena, T. Ganz, and E. Nemeth. 2020. "Maternal Hepcidin Suppression Is Essential for Healthy Pregnancy" American Society of Hematology Annual Meeting and Exposition, Virtual.

Schmidt, Paul J. 2015. "Regulation of Iron Metabolism by Hepcidin under Conditions of Inflammation." *Journal of Biological Chemistry 290* (31): 18975–18983.

Schuller, Kyla. 2018. *The Biopolitics of Feeling: Race, Sex, and Science in the Nineteenth Century*. Durham: Duke University Press.

Seyoum, Yohannes, Kaleab Baye, and Christèle Humblot. 2021. "Iron Homeostasis in Host and Gut Bacteria—a Complex Interrelationship." *Gut Microbes 13* (1): 1874855. https://doi.org/10.1080/19490976.2021.1874855.

Sheftel, Alex D., Anne B. Mason, and Prem Ponka. 2012. "The Long History of Iron in the Universe and in Health and Disease." *Biochimica et Biophysica Acta (BBA)-General Subjects 1820* (3): 161–187. https://doi.org/10.1016/j.bbagen.2011.08.002.

Shell-Duncan, Bettina, and Thomas McDade. 2005. "Cultural and Environmental Barriers to Adequate Iron Intake among Northern Kenyan Schoolchildren." *Food and Nutrition Bulletin 26* (1): 39–48.

Shell-Duncan, Bettina, and Stacie A Yung. 2004. "The Maternal Depletion Transition in Northern Kenya: The Effects of Settlement, Development and Disparity." *Social Science & Medicine 58* (12): 2485–2498.

Sloan, AW. 1987. "Thomas Sydenham, 1624–1689." *South African Medical Journal 72* (4): 275–278.

Slocum, Sally. 1975. "Woman the Gatherer: Male Bias in Anthropology." In *Toward an Anthropology of Women*, edited by Rayna R. Reiter, 36–50. New York: Monthly Reviews Press.

Smith, M. R., C. D. Golden, and S. S. Myers. 2017. "Potential Rise in Iron Deficiency Due to Future Anthropogenic Carbon Dioxide Emissions." *GeoHealth 1* (6): 248–257. https://doi.org/10.1002/2016GH000018.

Snook, Jonathon, Neeraj Bhala, Ian L. P. Beales, David Cannings, Chris Kightley, Robert P. H. Logan, D. Mark Pritchard, et al. 2021. "British Society of Gastroenterology Guidelines for the Management of Iron Deficiency Anaemia in Adults." *Gut 70* (11): 2030–2051. https://doi.org/10.1136/gutjnl-2021-325210.

Sousa Gerós, A., A. Simmons, H. Drakesmith, A. Aulicino, and J. N. Frost. 2020. "The Battle for Iron in Enteric Infections." *Immunology 161* (3): 186–199. https://doi.org/10.1111/imm.13236.

Stanford, Craig B., and Henry T. Bunn. 2001. *Meat-Eating and Human Evolution.* New York: Oxford University Press.

Starobinski, Jean. 1981. "Chlorosis–the 'Green Sickness.'" *Psychological Medicine, 11* (3): 459–468.

Steckel, Richard H., Jerome C. Rose, Clark Spencer Larsen, and Phillip L. Walker. 2002. "Skeletal Health in the Western Hemisphere from 4000 BC to the Present." *Evolutionary Anthropology 11* (4): 142–155.

Stoffel, N. U., C. Zeder, G. M. Brittenham, D. Moretti, and M. B. Zimmermann. 2020. "Iron Absorption from Supplements Is Greater with Alternate Day Than with Consecutive Day Dosing in Iron-Deficient Anemic Women." *Haematologica 105* (5): 1232–1239. https://doi.org/10.3324/haematol.2019.220830.

Stoltzfus, Rebecca J. 2003. "Iron Deficiency: Global Prevalence and Consequences." *Food and Nutrition Bulletin 24* (4_suppl_1): S99–S103.

Strange, Julie-Marie. 2000. "Menstrual Fictions: Languages of Medicine and Menstruation, c. 1850–1930." *Women's History Review 9* (3): 607–628.

Strassmann, Beverly I. 1996. "The Evolution of Endometrial Cycles and Menstruation." *The Quarterly Review of Biology 71* (2): 181–220.

Strassmann, Beverly I. 1997. "The Biology of Menstruation in *Homo sapiens*: Total Lifetime Menses, Fecundity, and Nonsynchrony in a Natural-Fertility Population." *Current Anthropology 38* (1): 123–129.

Sullivan, Kevin M., Zuguo Mei, Laurence Grummer-Strawn, and Ibrahim Parvanta. 2008. "Haemoglobin Adjustments to Define Anaemia." *Tropical Medicine & International Health 13* (10): 1267–1271. https://doi.org/10.1111/j.1365-3156.2008.02143.x.

Tan, Zhu-Xia, Rattan Lal, and Keith D. Wiebe. 2005. "Global Soil Nutrient Depletion and Yield Reduction." *Journal of Sustainable Agriculture 26* (1): 123–146.

Traeger, Lisa, Caroline A. Enns, Jan Krijt, and Andrea U. Steinbicker. 2018. "The Hemochromatosis Protein HFE Signals Predominantly via the BMP Type I Receptor ALK3 in vivo." *Communications Biology 1* (1): 65. https://doi.org/10.1038/s42003-018-0071-1.

Trevathan, Wenda R., and K. P. Rosenberg. 2020. "Evolutionary Medicine and Women's Reproductive Health." In *Integrating Evolutionary Biology into Medical Education: For Maternal and Child Healthcare Students, Clinicians, and Scientists*, edited by Jay Schulkin and Michael Power, 47–66. New York: Oxford University Press.

Trousseau, A. 1870. "Glycosuria: Saccharine Diabetes." In *Lectures on Clinical Medicine Delivered at the Hotel Dieu, Paris*, edited and translated by J. R. Cormack, 491–527. London: New Sydenham Society.

Tsing, Anna Lowenhaupt. 2015. *The Mushroom at the End of the World*. Princeton, NJ: Princeton University Press.

US National Institutes of Health Office of Dietary Supplements. 2022. "Iron: Fact sheet for health professionals." Accessed October 14, 2022. https://ods.od.nih.gov/factsheets/Iron-HealthProfessional/.

Valeggia, Claudia, and Peter T. Ellison. 2009. "Interactions between Metabolic and Reproductive Functions in the Resumption of Postpartum Fecundity." *American Journal of Human Biology 21* (4): 559–566.

VanderMeulen, H., R. Strauss, Y. Lin, A. McLeod, J. Barrett, M. Sholzberg, and J. Callum. 2020. "The Contribution of Iron Deficiency to the Risk of Peripartum Transfusion: A Retrospective Case Control Study." *BMC Pregnancy Childbirth 20* (1): 196. https://doi.org/10.1186/s12884-020-02886-z.

Venuti, Silvia, Laura Zanin, Fabio Marroni, Alessandro Franco, Michele Morgante, Roberto Pinton, and Nicola Tomasi. 2019. "Physiological and Transcriptomic Data Highlight Common Features between Iron and Phosphorus Acquisition Mechanisms in White Lupin Roots." *Plant Science 285*: 110–121. https://doi.org/https://doi.org/10.1016/j.plantsci.2019.04.026.

Verma, Smriti, and Bobby J. Cherayil. 2017. "Iron and Inflammation—the Gut Reaction." *Metallomics: Integrated Biometal Science 9* (2): 101–111. https://doi.org/10.1039/c6mt00282j.

Vonderheid, Susan C., Lisa Tussing-Humphreys, Chang Park, Heather Pauls, Nefertiti OjiNjideka Hemphill, Bazil LaBomascus, Andrew McLeod, and Mary Dawn Koenig. 2019. "A Systematic Review and Meta-Analysis on the Effects of Probiotic Species On Iron Absorption and Iron Status." *Nutrients 11* (12): 2938.

Wächtershäuser, Günter. 1988a. "Before Enzymes and Templates: Theory of Surface Metabolism." *Microbiological Reviews 52* (4): 452–484. https://doi.org/10.1128/mr.52.4.452-484.1988.

Wächtershäuser, Günter. 1988b. "Pyrite Formation, the First Energy Source for Life: A Hypothesis." *Systematic and Applied Microbiology 10* (3): 207–210. https://doi.org/https://doi.org/10.1016/S0723-2020(88)80001-8.

Wächtershäuser, Günter. 1992. "Groundworks for an Evolutionary Biochemistry: The Iron-Sulphur World." *Progress in Biophysics and Molecular Biology 58* (2): 85–201. https://doi.org/10.1016/0079-6107(92)90022-X.

Wade, Willoughby Francis. 1872. "Clinical Lecture on the Relation between Menstruation and the Chlorosis of Young Women." *British Medical Journal 2* (602): 35–37. https://doi.org/10.1136/bmj.2.602.35.

Wailoo, Keith. 1999. *Drawing Blood: Technology and Disease Identity in Twentieth-Century America*. Baltimore, MD: Johns Hopkins University Press.

Walker, P. L., R. R. Bathurst, R. Richman, T. Gjerdrum, and V. A. Andrushko. 2009. "The Causes of Porotic Hyperostosis and Cribra Orbitalia: A Reappraisal of the Iron-Deficiency-Anemia Hypothesis." *American Journal of Physical Anthropology 139* (2): 109–125. https://doi.org/10.1002/ajpa.21031.

Wallace, D. F., and V. N. Subramaniam. 2016. "The Global Prevalence of HFE and non-HFE Hemochromatosis Estimated from Analysis of Next-Generation Sequencing Data." *Genetic Medicine 18* (6): 618–626. https://doi.org/10.1038/gim.2015.140.

Wander, Katherine, Bettina Shell-Duncan, and Thomas W McDade. 2009. "Evaluation of Iron Deficiency as a Nutritional Adaptation to Infectious Disease: An Evolutionary Medicine Perspective." *American Journal of Human Biology 21* (2): 172–179.

Wassef, Andréanne, Quoc Dinh Nguyen, and Martin St-André. 2019. "Anaemia and Depletion of Iron Stores as Risk Factors for Postpartum Depression: A Literature Review." *Journal of Psychosomatic Obstetrics & Gynecology 40* (1): 19–28. https://doi.org/10.1080/0167482X.2018.1427725.

Weinberg, E. D. 1984. "Iron Withholding: A Defense against Infection and Neoplasia." *Physiological Reviews 64* (1): 65–102. https://doi.org/10.1152/physrev.1984.64.1.65.
Weinberg, ED. 2010. "Can Iron Be Teratogenic?" *Biometals 23* (2): 181–184.
Wiley, Andrea S., and Solomon H. Katz. 1998. "Geophagy in Pregnancy: A Test of a Hypothesis." *Current Anthropology 39* (4): 532–545.
Weiss, D. J., T. C. D. Lucas, M. Nguyen, A. K. Nandi, D. Bisanzio, K. E. Battle, E. Cameron, et al. 2019. "Mapping the Global Prevalence, Incidence, and Mortality of *Plasmodium falciparum*, 2000–17: A Spatial and Temporal Modelling Study." *The Lancet 394* (10195): 322–331. https://doi.org/10.1016/s0140-6736(19)31097-9.
Williams, Robert Joseph Paton. 1981. "The Bakerian Lecture, 1981: Natural Selection of the Chemical Elements." *Proceedings of the Royal Society of London. Series B. Biological Sciences 213* (1193): 361–397.
Wilunda, Calistus, Milkah Wanjohi, Risa Takahashi, Elizabeth Kimani-Murage, and Antonina Mutoro. 2022. "Association of Women's Empowerment with Anaemia and Haemoglobin Concentration in Children in sub-Saharan Africa: A Multilevel Analysis." *Maternal & Child Nutrition 19*: e13426. https://doi.org/10.1111/mcn.13426.
World Bank. 2021. "Mortality Rate, Adult, Female (per 1,000 Female Adults)—Kenya, United States." Accessed August 31, 2021. https://data.worldbank.org/indicator/SP.DYN.AMRT.FE?locations=KE-US.
World Bank. 2022. "Poverty." Accessed October 5, 2022. https://www.worldbank.org/en/topic/poverty.
World Health Organization. 2016. *Guideline Daily Iron Supplementation in Infants and Children*. Geneva: World Health Organization.
World Health Organization. 2022a. "Fact Sheet: Malaria." October 6, 2022. https://www.who.int/news-room/fact-sheets/detail/malaria.
World Health Organization. 2022b. "Soil-Transmitted Helminth Infections." Accessed October 6, 2022. https://www.who.int/news-room/fact-sheets/detail/soil-transmitted-helminth-infections.
Xu, P., F. Hong, J. Wang, Y. Cong, S. Dai, S. Wang, J. Wang, et al. 2017. "Microbiome Remodeling via the Montmorillonite Adsorption-Excretion Axis Prevents Obesity-Related Metabolic Disorders." *EBioMedicine 16*: 251–261. https://doi.org/10.1016/j.ebiom.2017.01.019.
Young, Sera L. 2011. *Craving Earth*. New York: Columbia University Press.
Young, Sera L., Paul W. Sherman, Julius B. Lucks, and Gretel H. Pelto. 2011. "Why on Earth?: Evaluating Hypotheses about the Physiological Functions of Human Geophagy." *The Quarterly Review of Biology 86* (2): 97–120.
Young, Sera L., M. Jeffrey Wilson, Stephen Hillier, Evelyne Delbos, Said M. Ali, and Rebecca J. Stoltzfus. 2010. "Differences and Commonalities in Physical, Chemical and Mineralogical Properties of Zanzibari Geophagic Soils." *Journal of Chemical Ecology 36* (1): 129–140.
Zimmermann, Michael B., Christophe Chassard, Fabian Rohner, Eliézer K. N'Goran, Charlemagne Nindjin, Alexandra Dostal, Jürg Utzinger, et al. 2010. "The Effects of Iron Fortification on the Gut Microbiota in African Children: A Randomized Controlled Trial in Côte d'Ivoire." *The American Journal of Clinical Nutrition 92* (6): 1406–1415. https://doi.org/10.3945/ajcn.110.004564.
Zimmermann, Michael B., and Richard F Hurrell. 2007. "Nutritional Iron Deficiency." *The Lancet 370* (9586): 511–520.

Index

For the benefit of digital users, indexed terms that span two pages (e.g., 52–53) may, on occasion, appear on only one of those pages.

Tables and figures are indicated by *t* and *f* following the page number